# Lecture Notes in Computer Science 6380

*Commenced Publication in 1973*
Founding and Former Series Editors:
Gerhard Goos, Juris Hartmanis, and Jan van Leeuwen

W0192947

Abdelkader Hameurlain   Josef Küng
Roland Wagner   Torben Bach Pedersen
A Min Tjoa (Eds.)

# Transactions on Large-Scale Data- and Knowledge-Centered Systems II

 Springer

Editors-in-Chief

Abdelkader Hameurlain
Paul Sabatier University
Institut de Recherche en Informatique de Toulouse (IRIT)
118, route de Narbonne, 31062 Toulouse Cedex, France
E-mail: hameur@irit.fr

Josef Küng
Roland Wagner
University of Linz, FAW
Altenbergerstraße 69
4040 Linz, Austria
E-mail: {jkueng,rrwagner}@faw.at

Guest Editors

Torben Bach Pedersen
Aalborg University
Department of Computer Science
Selma Lagerløfs Vej 300
9220 Aalborg, Denmark
E-mail: tbp@cs.aau.dk

A Min Tjoa
Vienna University of Technology
Institute of Software Technology
Favoritenstr. 9-11/188
1040 Vienna, Austria
E-mail: amin@ifs.tuwien.ac.at

Library of Congress Control Number: 2009932361

CR Subject Classification (1998): H.2, I.2.4, H.3, H.4, J.1, H.2.8

ISSN      0302-9743 (Lecture Notes in Computer Science)
ISSN      1869-1994 (Trans. on Large-Scale Data- and Knowledge-Centered Systems)
ISBN-10   3-642-16174-X Springer Berlin Heidelberg New York
ISBN-13   978-3-642-16174-2 Springer Berlin Heidelberg New York

springer.com

© Springer-Verlag Berlin Heidelberg 2010
Printed in Germany

Typesetting: Camera-ready by author, data conversion by Scientific Publishing Services, Chennai, India
Printed on acid-free paper      06/3180

# Preface

This special issue of TLDKS contains two kinds of papers. First, it contains a selection of the best papers from the 11th International Conference on Data Warehousing and Knowledge Discovery (DaWaK 2009), which was held from August 31 to September 2, 2009 in Linz, Austria. Second, it contains a special section of papers on a particularly challenging domain in information retrieval, namely patent retrieval.

Over the last decade, the International Conference on Data Warehousing and Knowledge Discovery (DaWaK) has established itself as one of the most important international scientific events within data warehousing and knowledge discovery. DaWaK brings together a wide range of researchers and practitioners working on these topics. The DaWaK conference series thus serves as a leading forum for discussing novel research results and experiences within the field. The 11th International Conference on Data Warehousing and Knowledge Discovery (DaWaK 2009) continued the tradition by disseminating and discussing innovative models, methods, algorithms, and solutions to the challenges faced by data warehousing and knowledge discovery technologies.

The papers presented at DaWaK 2009 covered a wide range of issues within data warehousing and knowledge discovery. Within data warehousing and analytical processing, the topics covered were: data warehouse modeling including advanced issues such as spatio-temporal warehouses and DW security, OLAP on data streams, physical design of data warehouses, storage and query processing for data cubes, advanced analytics functionality, and OLAP recommendation. Within knowledge discovery and data mining, the topics included: stream mining, pattern mining for advanced types of patterns, advanced rule mining issues, advanced clustering techniques, spatio-temporal data mining, data mining applications, as well as a number of advanced data mining techniques. It was encouraging to see that many papers covered important emerging issues such as: spatio-temporal data, streaming data, non-standard pattern types, advanced types of data cubes, complex analytical functionality including recommendations, multimedia data, mssing and noisy data, as well as real-world applications within genes, and the clothing and telecom industries. The wide range of topics bears witness to the fact that the data warehousing and knowledge discovery field is dynamically responding to the new challenges posed by novel types of data and applications.

The DaWaK 2009 Call for Papers attracted a large number of quality submissions. From 124 submitted abstracts, we received 100 papers from 17 countries in Europe, North America, and Asia. The program committee finally selected 36 papers, yielding an acceptance rate of 36%. The DaWaK proceedings were published by Springer in the LNCS series.

A few of the papers were selected by the DaWaK PC chairs based on both quality and potential, and the authors were invited to submit significantly extended versions for a new round of reviewing. After a thorough refereeing process, including further revisions of the papers, the following three papers were finally accepted for inclusion in this special issue of TLDKS.

In the paper "Fast Loads and Fast Queries", Graefe and Kuno aim at the double goal of achieving fast bulk loading, while still supporting fast queries through computing redundant search structures on the fly. The paper introduces a range of new techniques to achieve this, named zone filters, zone indexes, adaptive merging, and partition filters, respectively. These techniques address the limitations of earlier techniques like Netezza zone maps without reducing the advantages. The proposed data structures can be created on the fly, as a "side effect" of the load process. All required analyses can be performed with a moderate amount of new data in the main memory buffer pool, and traditional sorting and indexing are not required. However, the observed query performance is as good as that of zone maps where those can be used, and is better than that of zone maps for the (many) predicates where zone maps are ineffective. Finally, simulations show that the query performance is comparable to query processing in a database with traditional indexing, but without the long loading time that such indexing requires.

In the paper "Discovery of Frequent Patterns in Transactional Data Streams", Ng and Dash investigate the problem of discovering frequent patterns in a transaction stream. They first survey two common methods from the literature: 1) approximate counting, e.g., lossy counting (LCA), using a lower support threshold and 2) maintaining a running sample, e.g., reservoir sampling (Algo-Z), and generating frequent itemsets on demand from this sample. Then, the pros and cons of each method are discussed. The authors then propose a novel sampling algorithm, called DSS, which selects a transaction to include in the sample based on single item histograms. Finally, an experimental comparison between the three approaches is performed, showing that DSS is the most accurate, then LCA, then Algo-Z, while Algo-Z is the fastest, followed by DSS, and then LCA.

In the paper "Efficient Online Aggregates in Dense-Region-Based Data Cube Representations", Haddadin and Lauer investigate the space- and time-efficient in-memory representation of large data cubes. The proposed solution builds on the well known idea of identifying dense sub-regions of the cube and storing dense and sparse regions separately, but improves upon it by focusing not only on space- but also on time-efficiency. This is done both for the initial dense-region extraction and for queries carried out in the resulting hybrid data structure. The paper first describes a pre-aggregation method for representing dense sub-cubes which supports efficient online aggregate queries as well as cell updates. Then, the sub-cube extraction approach is presented. Two optimization methods that trade available memory for increased aggregate query performance are presented, along with the adaptation to multi-core/multi-processor architectures. Experiments with several real-world data are performed, showing how the algorithms can be tuned by setting a number of parameters.

Moving from the analysis of structured data to a semi-structured domain, patent retrieval poses a range of demanding challenges from an information retrieval perspective. Part of this is due to the sheer volume of data, part to the rather specific requirements on the retrieval scenarios. Patent retrieval, for example, is one of the few settings where recall is commonly more important than precision. Different to normal web retrieval, where the goal is to provide a set of relevant documents within the top-10 documents retrieved, patent retrieval requires all documents to be found, with result lists being scanned ranging well into the hundreds.

The two papers selected for this special section address a specific characteristic of this domain, namely the question of whether a system is able to find specific documents at all. As only a limited number of documents can be examined in any setting, there may be documents that do not appear within the top-n ranks for any realistic query, thus remaining practically irretrievable. Both papers in this special section address this retrievability challenge from two different perspectives.

The paper "Improving Access to Large Patent Corpora" by Bache and Azzopardi analyzes the characteristics of the retrieval system itself, measuring the system bias towards specific document characteristics. Specifically, a system's sensitivity to term frequency characteristics of the documents, length variations and convexity, i.e., managing the impact of phrases, are evaluated. It then proposes a hybrid retrieval system, combining both exact as well as best-match principles.

The paper "Improving Retrievability and Recall by Automatic Corpus Partitioning" by Bashir and Rauber aims to overcome the retrievability challenge by automatically classifying documents into ones with potentially high and low retrievability based on statistical characteristics of word distributions. Using a split corpus, separate retrieval engines can be applied, allowing documents from the lower retrievability section of a document corpus to find their way into the result set, thus ensuring better coverage and mitigating the risk of having un-findable documents.

As the Guest Editors of this special issue, we would like to thank all the referees, both the DaWaK 2009 PC members and the extra reviewers for the special issue, for their careful and dedicated work. We hope you will enjoy the papers that follow and see them as bearing witness to the high quality of the DaWaK conference series as well as to the specific challenges faced by retrieval engines in highly specialized domains of large volume data.

July 2010

Torben Bach Pedersen
A Min Tjoa

# Editorial Board

# Table of Contents

## Data Warehousing and Knowledge Discovery

## Information Retrieval

# Discovery of Frequent Patterns in Transactional Data Streams

Willie Ng and Manoranjan Dash

Centre for Advanced Information Systems,
Nanyang Technological University,
Singapore 639798
{ps7514253f,AsmDash}@ntu.edu.sg

**Abstract.** A data stream is generated continuously in a dynamic environment with huge volume, infinite flow, and fast changing behaviors. There have been increasing demands for developing novel techniques that are able to discover interesting patterns from data streams while they work within system resource constraints. In this paper, we overview the state-of-art techniques to mine frequent patterns in a continuous stream of transactions. In the literature two prominent approaches are often used: (a) perform approximate counting (e.g., lossy counting algorithm (LCA) of Manku and Motwani, VLDB 2002) by using a lower support threshold than the one given by the user, or (b) maintain a running sample (e.g., reservoir sampling (Algo-Z) of Vitter, TOMS 1985) and generate frequent patterns from the sample on demand. Although both approaches are practically useful, to the best of our knowledge there has been no comparison between the two approaches. We also introduce a novel sampling algorithm (*DSS*). *DSS* selects transactions to be included in the sample based on histogram of single itemsets. An empirical comparison study between the 3 algorithms is performed using synthetic and benchmark datasets. Results show that *DSS* is consistently more accurate than LCA and Algo-Z, whereas LCA performs consistently better than Algo-Z. Furthermore, *DSS*, although requires more time than Algo-Z, is faster than LCA.

## 1 Introduction

In this paper, we focus on frequent pattern mining (FPM) over streaming data. The problem of mining frequent patterns online can be formally stated as follows: A transaction contains a set of items. Let $\mathcal{I} = \{i_1, i_2, \ldots, i_n\}$ be a set of distinct literals, called items. An itemset $X$ is an arbitrary set of items. The frequency of an itemset $X$, denoted by $freq(X)$, is the number of transactions in the data stream $Ds$ that contain $X$. The support of an itemset $X$, denoted by $\sigma(X)$, is the ratio of the frequency of $X$ to the number of transactions processed so far $N$, i.e., $\sigma(X) = freq(X)/N$. Given a pre-defined support threshold $\sigma_{min}$, an itemset $X$ is considered a frequent pattern if its frequency is more than or equal to $\sigma_{min} \times N$. We need to find all frequent patterns.

FPM is extremely popular particularly amongst researchers of data mining. On static data, many algorithms on FPM have been proposed. This research has led to further efforts in various directions [HCXY07]. But for streaming data the advancement in

A. Hameurlain et al. (Eds.): TLDKS II, LNCS 6380, pp. 1–30, 2010.

research has not been so spectacular. Although for several years many researchers have been proposing algorithms on FPM over streaming data (first prominent paper appeared in VLDB 2002 [MM02][1]), even the recent papers on the topic [CDG07] show that FPM in streaming data is not trivial. Manku and Motwani nicely described this problem in their seminal work - Lossy Counting Algorithm (LCA) [MM02]. Their work led to many other similar papers that use approximate counting [CKN08, CKN06, GHP+03].

At the heart of approximate counting is the fact that for stream data one cannot keep exact frequency count for all possible itemsets. Note that here we are concerned with datasets that have 1000s of items or more, not just toy datasets with less than 10 items. The power set of the set of items cannot be maintained in today's memory. To solve this memory problem, LCA maintains only those itemsets which are frequent in at least a small portion of the stream, but if the itemset is found to be infrequent it is discontinued. As LCA does not maintain any information about the stream that has passed by, it adds an error frequency term to make the total frequency of a "potentially" frequent itemset higher than its actual frequency. Thus, LCA produces 100% recall but suffers from poor precision. These facts are described in [ND08].

Sampling is another approach that is used in [Toi96, MTV94, ZPLO96, YCLZ04] to produce frequent itemsets. The idea is very simple: maintain a sample over the stream and when asked, run the frequent pattern mining algorithm, such as the Apriori algorithm, to output the frequent itemsets. Research in sampling focuses on how to maintain a sample over the streaming data. On one hand approximate counting such as LCA does not keep any information about the stream that has passed by but keeps exact information starting from some point in the stream. On the other hand sampling keeps information about the whole stream (one can use a decaying factor to decrease or increase the influence of the past) but only partially. So, unlike approximate counting, sampling will have both false positives and false negatives. A researcher would love to know how these two compare against each other. Later, we will give an overview of these two approaches.

We have an additional contribution in this paper. Simple random sampling (SRS), or its counterpart reservoir sampling in streaming data, is known to suffer from a few limitations: First, an SRS sample may not adequately represent the entire data set due to random fluctuations in the sampling process. This difficulty is particularly apparent at small sample ratios which is the case for very large databases with limited memory. Second, SRS is blind towards noisy data objects, i.e., it treats both *bona fide* and noisy data objects similarly. The proportion of noise in the SRS sample and the original data set are almost equal. So, in the presence of noise, performance of SRS degrades.

In this paper a new sampling algorithm called *DSS* (Distance based Sampling for Streaming data) is proposed for streaming data. The main contributions of this proposed method are: *DSS* elegantly addresses both issues (very large size of the data or in other words, very small sampling ratio, and noise) simultaneously. Experiments in FPM are conducted, and in all three of them the results show convincingly that *DSS* is far superior to *SRS* at the expense of a slight amount of processing time. Later in the conclusion the trade-off analysis between accuracy and time shows that it is worthwhile to invest in

---

[1] Prior to this work, Misra and Gries [MG82] proposed the deterministic algorithm for $\epsilon$-approximate frequency counts. The same algorithm has been rediscovered recently by Karp et al. [KSP03].

*DSS* than other approaches including *SRS* particularly when the domain is very large and noisy. Experiments are done mainly using synthetic datasets from IBM QUEST project and benchmark dataset, Kosarak.

## 2   Review 1 : Approximate Counting

A transaction data stream is a sequence of incoming transactions, $Ds = (t_1, t_2, ..., t_i, ...)$, where $t_i$ is the $i$-th transaction. Here, a portion of the stream is called a window. A window, $W$, is a subsequence of the stream between the $i$th and the $j$th point where $W = (T_i, T_{i+1}, ..., T_j)$ and $i < j$. The point, $T$, can be either time-based or count-based. For time-base, $W$ consists of a sequence of fixed length time units, where a variable number of transactions may arrive within each time unit. As for count-based, $W$ is composed of a sequence of batches, where each batch consists of an equal number of transactions. To process and mine the stream, different window models are often employed. In this paper, we identify four types of data stream mining models: *landmark window, sliding window, damped window* and *tilted-time window* [Sha02, JG06, Agg07].

### 2.1   Landmark Window

A model is considered a *landmark window* if $W = (T_i, T_{i+1}, ..., T_\tau)$, where $\tau$ is the current time point. In this model, we are mining for frequent patterns starting from the time point called landmark, $i$, to the most recent time point (see Figure 1). Usually, $i$ is set to 1 and thus we are trying to discover frequent patterns over the entire history of the data stream.

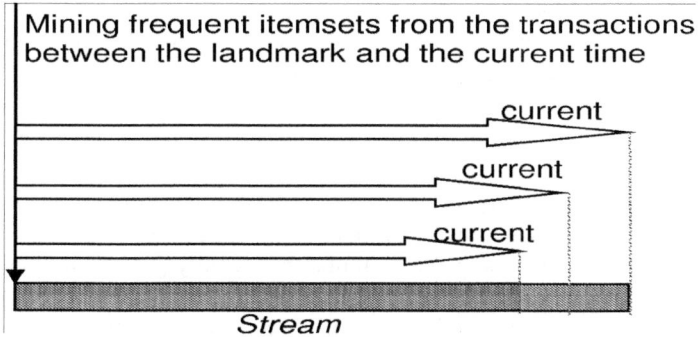

**Fig. 1.** Landmark model

Manku and Motwani [MM02] proposed a method for mining all frequent itemsets over the life of the stream. LCA stores itemsets in a tree structure. With each itemset stored, they record an under-estimate of the frequency and an error term. Upon the arrival of a new batch of transactions, the tree is updated; some itemsets may be dropped and others may be added. The dropping and adding conditions ensure that: (1) the estimated frequency of all itemsets in the tree are less than the true frequency by no more than $\epsilon N$ (N is the length of the stream and $\epsilon$ is a user-defined error bound) and (2) no frequent itemset is omitted from the tree. By scanning the tree and returning all patterns whose estimated frequency is greater than $(\sigma_{min} - \epsilon) \times N$, an approximation to the frequent patterns can be made. For single itemset mining, LCA requires at most $\frac{1}{\epsilon} log(\epsilon N)$ of memory space.

Similarly, Li et al. proposed the online incremental projected, single-pass algorithms, called DSM-FI [LLS04] and DSM-MFI [LLS05] to mine the set of all frequent itemsets and maximal frequent itemsets over the entire history of data streams. Extended prefix-tree based data structures are developed in the proposed algorithms to store items and their support values, window ids, and nodes links pointing to the root or a certain node. The experiments show that when data set is dense, DSM-FI is more efficient than LCA.

Observe that in a landmark window, we consider all the time points from the landmark till the most recent point as equally important. There is no distinction made between past and present data. However, since data streams may contain time-varying data elements, patterns which are frequent may evolve as well. Often these changes make the mining result built on the old transactions inconsistent with the new ones. To emphasize the recency of data, we can adopt the *sliding window, damped window* or *tilted-time window*.

## 2.2   Sliding Window

As a simple solution for time changing environment, it is possible to consider the *sliding window* model (see Figure 2). A *sliding window* keeps a window of size $w$. Here, we are interested to mine the frequent patterns in the window $W = (T_{\tau-w+1}, ..., T_{\tau})$ where $T_i$ is a batch or time unit. When a new transaction arrives, the oldest resident in the window is considered obsolete and deleted to make room for the new one [BDM02]. The mining result is totally dependent on the range of the window. Therefore, all the transactions in the window need to be maintained in order to remove their effects on the current mining result when they become obsolete.

Based on the estimation mechanism of LCA, Chang and Lee introduced a sliding window method for finding recent frequent patterns in a data stream when the minimum support threshold, error parameter and sliding window size are provided [CL04]. A recent frequent pattern is one whose support in the transactions within the current sliding window is greater than or equal to the support threshold. By restricting the target of a mining process as a fixed number of recently generated transactions in a data stream, the recent change of the data stream can be efficiently analyzed. Consequently, the proposed method can catch the recent change of a data stream as well as analyze the past variations of knowledge embedded in a data stream. In addition, Chang and Lee also proposed another similar algorithm $(estWin)$ to maintain frequent patterns over a sliding window [CL03a].

Chi et al. proposed a new mining algorithm, called MOMENT to monitor transactions in the sliding window so that we can output the current closed frequent itemsets at any time [CWYM04]. Key to this algorithm is a compact data structure, the closed enumeration tree (CET), used to maintain a dynamically selected set of patterns over a sliding-window. The selected patterns consist of a boundary between closed frequent itemsets and the rest of the itemsets. Concept drifts in a data stream are reflected by

Starting Point

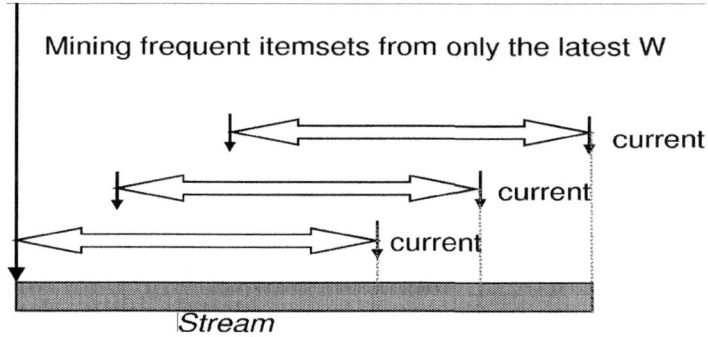

**Fig. 2.** Sliding-window model

boundary movements in the CET. An itemset that is frequent currently may not be frequent later. A status change of such itemset must occur through the boundary. Since the boundary is relatively stable, the cost of mining closed frequent itemsets over a sliding window is dramatically reduced to that of mining transactions that can possibly cause boundary movements in the CET.

In most algorithms that adopt the *sliding window* model, the biggest challenge will be quantifying $w$. In most cases, it is externally (user-)defined. However, if $w$ is too large and there is concept drift, it is possible to contain outdated information, which will reduce the learned model accuracy. On the other hand, if $w$ is too small, it may have insufficient data and the learned model will likely incur a large variance. Here, the open direction of research is to design a window that can dynamically decide its width based on the rate of the underlying change phenomenon [CDG07].

In addition, the use of a sliding window to mine for frequent patterns from the immediately preceding points may represent another extreme and rather unstable solution. This is because one may not wish to completely lose the entire history of the past data. Distant historical behavior and trend may also be queried periodically. In such cases, we can adopt the *damped window* which does not erase old data completely but rather slowly "forgets" them.

### 2.3 Damped Window

The *damped window* (also called *time-fading window*) assigns different weights to transactions such that new transactions have higher weights than old ones [CL03b]. The use of the *damped window* diminishes the effect of the old and obsolete information of a data stream on the mining result. At every moment, based on a fixed decay rate [CS03] or forgetting factor [YSJ+00], a transaction processed $d$ time steps ago is assigned a weight $m^d$, where $m < 1$. The weight decreases as time passes by (see Figure 3). In general, the closer to 1 the decay, the more the history is taken into account. Correspondingly, the count of an itemset is also defined based on the weight of each transaction.

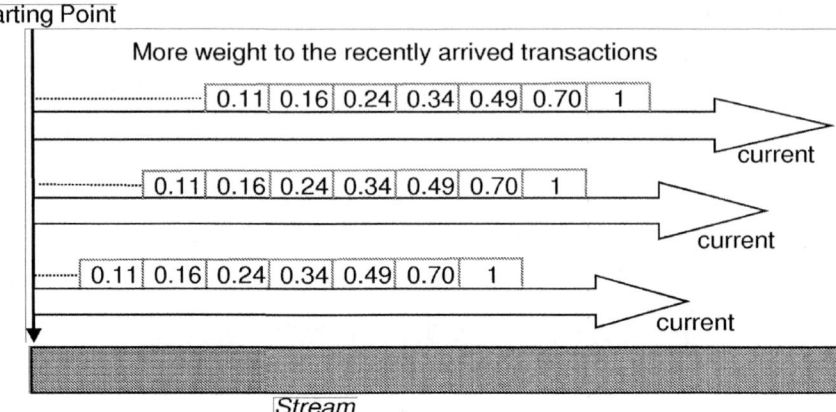

**Fig. 3.** Time-fading model

Chang and Lee proposed an *estDec* method for mining recent frequent patterns adaptively over an online data stream [CL03b]. *estDec* examines each transaction in a data stream one-by-one without candidate generation. The algorithm only needs to update the counts for the itemsets which are the subsets of the newly arrived transaction. The occurrence count of a significant pattern that appears in each transaction is maintained by a prefix-tree lattice structure in the main memory. The effect of old transactions on the current mining result is diminished by decaying the old occurrence count of each pattern as time goes by. In addition, the rate of decay of old information is flexibly defined as needed. The total number of significant patterns in the main memory is minimized by delayed-insertions and pruning operations on an itemset. As a result, its processing time is flexibly governed while sacrificing its accuracy.

As highlighted in [CKN08], estimating the frequency of a pattern from the frequency of its subsets can generate a large error. The error may propagate all the way from 2-subsets to n-supersets. Therefore, it is difficult to formulate an error bound on the computed frequency of the resulting itemsets and a large number of the false positive itemsets will be produced, since the computed frequency of an itemset can be larger than its actual frequency. Moreover, *damped* model suffers from similar problems as in *sliding windows* model. The challenge of determining a suitable $w$ in sliding windows model is now translated to that of determining a suitable $m$.

### 2.4 Tilted-Time Window

In data stream analysis, some people are more interested in recent changes at a fine scale than in long term changes at a coarse scale. We can register time at different levels of granularity. The most recent time is registered at the finest granularity while the earlier time is registered at a coarser granularity. Such time dimension model is often called a tilted-time frame [CDH+02]. In this model, we are interested in frequent patterns over a set of windows. Here, each window corresponds to different time granularity based on their recency. There are several ways to design a titled-time frame. For example,

Figure 4 shows such a *tilted-time window*: the most recent 4 quarters of an hour, then the last 24 hours, and 31 days. Based on this model, one can compute frequent patterns in the last hour with the precision of quarter of an hour, the last day with the precision of hour, and so on, until the whole month. This model registers only $4 + 24 + 31 = 59$ units of time, with an acceptable trade-off of lower granularity at a distance time (see Figure 4). With this model, we can compute frequent patterns with different precisions.

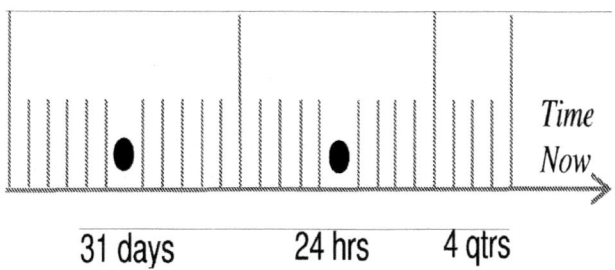

**Fig. 4.** Tilted-time model

Giannella et.al. make use of the characteristics of the FP-Growth [HPY00] and a logarithmic *tilted-time window* to mine frequent patterns in data stream. Their algorithm, FP-Stream, is able to count itemsets and save the frequent patterns in an effective FP-Tree-based model [GHP+03]. FP-Stream can be seen as a combination of the different models. For an itemset, frequencies are computed for the most recent windows of lengths $w$, $2w$, $4w$, $8w$, etc. So, the more recent part of the stream is covered more thoroughly. The combination of these frequencies permit efficient query answering over the history of the stream.

Although this method adopts the *tilted-time window* to solve the aforesaid problem in different granularity and to update frequent patterns with the incoming data stream, it considers all the transactions as the same. In the long run, FP-Stream will inevitably become very large over time and updating and scanning such a large structure may degrade its performance. In this respect, we need to design a better data structure to mine frequent patterns efficiently.

## 2.5  Discussion

We have covered four different types of mining models. All these four models have been considered in current research on data stream mining. Choice of a model largely depends on the application needs. In this work, we consider the *landmark window*, finding frequent patterns over the entire data stream, as the most challenging and fundamentally important one. Often, it can serve as the basis to solve the latter three. For example, it is not difficult to see that an algorithm based on the *landmark window* can be converted to that using the *damped window* by adding a decay function on the upcoming data streams. It can also be converted to that using *sliding window* by keeping track of and processing data within a specified *sliding window*.

We note that all the previously described algorithms adopt a false-positive approach[2] (we will describe the false negative approach[3] in the next section). Such approach uses a relaxed minimum support threshold, $\epsilon$, to reduce the number of false positives and to obtain a more accurate frequency of the result itemsets. However, in reality, it is difficult to define $\epsilon$. Setting a wrong value for $\epsilon$ may either trigger a massive generation of false positives thus degrading the mining result, or slow down the processing speed due to over-intensive updating and scanning of the large data structure (see [ND08] for more details).

In addition, we observe that current data stream mining algorithms require the user to define one or more parameters before execution. Unfortunately, most of them do not provide any clue as to how we can adjust these parameters online while they are running. Some proposed methods let users adjust only certain parameters online, but these parameters may not be the key ones to the mining algorithms, and thus are not enough for a user friendly mining environment. From a practical point of view, it is also not feasible for user to wait for a mining algorithm stops before he can reset the parameters. This is because it may take a long time (or never) for the algorithm to finish due to the continuous arrivals and huge amount of data streams. For further improvement, we may consider either to allow the user to adjust online or to let the mining algorithm auto-adjust most of the key parameters in FPM, such as support, $\epsilon$, window size and decay rate.

## 3   Review 2 : Sampling

FPM algorithms require multiple passes over the whole database, and consequently the database size is by far the most influential factor of the execution time for very large databases. This is prohibitively expensive. For most mining tasks, exact counts are not necessarily required and an approximate representation is often more appropriate. This motivates an approach called data reduction (or synopsis representation) [BDF+97]. The problem of synopsis maintenance has been studied in great detail because of its applications to problems such as query estimation in data streams [AY06]. Several synopsis approaches such as sampling, histogram, sketches and wavelets are designed for use with specific applications. We do not attempt to survey all data reduction methods here. In this paper, we focus our discussion on sampling.

Sampling has great appeal because of its simplicity and general purpose reduction method which applies to a wide range of applications. Moreover, the benefits of sampling over other data reduction methods are more pronounced with increasing dimensionality: the larger the dimensionality, the more compact sampling becomes compared to, for example , multi-dimensional histograms or wavelet decompositions [BDF+97]. Also, sampling retains the relations and correlations between the dimensions, which may be lost by histograms or other reduction techniques. This latter point is important for data mining and analysis. Thus, the generated sample can be easily fed to arbitrary

---

[2] A false positive approach is one that returns a set of patterns that contains all frequent patterns and also some infrequent patterns.

[3] A false negative approach is one that returns a set of patterns that does not contain any infrequent patterns but may miss some frequent patterns.

data mining applications with little changes to the underlying method and algorithms. The survey by Olken and Rotem offers a good overview of sampling algorithms in databases [OR95]. Interested readers can also refer to [KK06] for a recent survey on association rule mining where some additional techniques on the use of sampling are given.

In recent years, a lot of algorithms based on sampling in mining frequent patterns in large databases have been proposed. In [CHS02], a method called *FAST* is introduced. It is a two-phase sampling based algorithm for discovering association rules in large databases. In Phase I, a large initial large sample of transactions is selected. It is used to quickly and almost accurately estimate the support of each individual item in the database. In Phase II, these estimated supports are used to either trim outlier transactions or select representative transactions from the initial sample, thereby forming a small final sample that more accurately reflects the statistical characteristics (i.e., itemset supports) of the entire database. The expensive operation of discovering all association rules is then performed on the final sample. In [HK06b], a new association mining algorithm is proposed that uses two phase sampling that addresses a limitation of *FAST*. *FAST* has the limitation that it only considers the frequent 1-itemsets in trimming/growing, thus, it did not have ways of considering mulit-itemsets including 2-itemsets. The new algorithm reflects the multi-itemsets in sampling transactions. It improves the mining results by adjusting the counts of both missing itemsets and false itemsets.

Lee et al. studied the usefulness of sampling techniques in data mining [LCK98]. By applying sampling techniques, they devised a new algorithm DELI for finding an approximate upper bound on the size of the difference between association rules of a database before and after it is updated. The algorithm uses sampling and statistical methods to give a reliable upper bound. If the bound is low, then the changes in association rules are small. So, the old association rules can be taken as a good approximation of the new ones. If the bound is high, the necessity of updating the association rules in the database is signaled. Experiments show that DELI is not only efficient with high reliability, but also scalable. It is effective in saving machine resources in the maintenance of association rules.

### 3.1  $\epsilon$-Approximation

The concept of $\epsilon$-approximation has been widely used in machine learning. A seminal result of Vapnik and Chervonenkis shows that a random sample of size $O(\frac{d}{\epsilon^2} \log \frac{1}{\epsilon})$, where $d$ is the "VC dimension" (assumed finite) of the data set, is an $\epsilon$ approximation. This result establishes the link between random samples and frequency estimations over several items simultaneously.

An example of $\epsilon$-approximation algorithm is the Sticky Sampling [MM02]. The algorithm is probabilistic in nature and can be employed to handle data streams. The user needs to specify three parameters: support $\sigma_{min}$, error $\epsilon$ and probability of failure $\delta$. Sticky Sampling uses a fixed size buffer and a variable sampling rate to estimate the counts of incoming items. The data structure $S$ is a set of entries of the form $(e, f)$, where $f$ estimates the frequency of an item $e$ belonging to the stream. Initially, $S$ is empty, and the sampling rate $r$ is set to 1. Sampling an item with rate $r$ means that the item is selected with probability $\frac{1}{r}$. For each incoming item $e$, if an entry for $e$ already exists in $S$, the corresponding frequency $f$ will be incremented; otherwise, sample the item with $r$. If this item is selected by sampling, we add an entry $(e, 1)$ to $S$ ; otherwise, ignore $e$ and move on to the next incoming item in the stream. Sticky Sampling is able to compute an $\epsilon$-deficient synopsis with probability at least $1 - \delta$ using at most $\frac{2}{\epsilon}\log(S^{-1}\delta^{-1})$ expected number of entries. While the algorithm can accurately maintain the statistics of items over stable data streams, it fails to address the needs of important applications, such as network traffic control and pricing, that require information about the entire stream but with emphasis on the most recent data. Moreover, Sticky Sampling cannot be applied in FPM because it only works well for frequent single itemsets.

In [BCD+03], an algorithm called *EASE* is proposed. *EASE* can deterministically produce a sample by repeatedly halving the data to arrive at the given sample size. The sample is produced by $\epsilon$-approximation method. *EASE* leads to much better estimation of frequencies than simple random sampling. Unlike *FAST* [CHS02], *EASE* provides a guaranteed upper bound on the distance between the initial sample and final sub-sample. In addition, *EASE* can process transactions on the fly, i.e., a transaction needs to be examined only once to determine whether it belongs to the final sub-sample. Moreover, the average time needed to process a transaction is proportional to the number of items in that transaction.

## 3.2 Progressive Sampling

Choosing a random sample of the data is one of the most natural ways of choosing a representative subset. The primary challenge in developing sampling-based algorithms stems from the fact that the support value of an itemset in a sample almost always deviates from the support value in the entire database. A wrong sample size may result in missing itemsets that have high support value in the database but not in the sample and false itemsets that are considered frequent in the sample but not in the database. Therefore, the determination of the sample size then becomes the critical factor in order to ensure that the outcomes of the mining process on the sample generalize well. However, the correct sample size is rarely obvious. How much sample is enough really depends on the combination of chosen mining algorithms, data set and application related loss function.

Parthasarathy [Par02] attempted to determine the sample size using progressive sampling [PJO99, ND06]. Progressive sampling is a technique that starts with a small sample, and then increases the sample size until a sample of sufficient size has been obtained. While this technique eliminates the need to determine the correct size initially, it requires that there be a way to evaluate the sample to judge if it is large enough. Parthasarathy's approach relies on a novel measure of model accuracy (self-similarity of associations across progressive samples), the identification of a representative class of frequent patterns that mimics the self-similarity values across the entire set of associations, and an efficient sampling methodology that hides the overhead of obtaining progressive samples by overlapping it with useful computation. However, in streaming data context, computing the sufficiency amount can be formidably expensive especially for hill-climbing based methods.

## 3.3    Statistical Sampling

We say that a data sample is sufficient if the observed statistics, such as mean or sample total have a variance lower than the predefined limits with high confidence. Large deviation bounds such as the Hoeffding bound (sometimes called the additive Chernoff bounds) can often be used to compute a sample size $n$ so that a given statistics on the sample is no more than $\varepsilon$ away from the same statistics on the entire database, where $\varepsilon$ is a tolerated error. As an example, if $\mu$ is the expected value of some amount in [0, 1] taken over all items in the database, and if $\hat{\mu}$ is a random variable denoting the value of that average over an observed sample, then the Hoeffding bound states that with probability at least $1 - \delta$,

$$|\mu - \hat{\mu}| \le \varepsilon. \tag{1}$$

The sample size must be at least $n$, where

$$n = \frac{1}{2\varepsilon^2} \cdot \ln \frac{2}{\delta}. \tag{2}$$

This equation is very useful and it has been successfully applied in classification as well as association rule mining [DH00, Toi96]. Although the Hoeffding bound is often considered conservative, it gives the user at least a certain degree of confidence. Note that demanding small error is expensive, due to the quadratic dependence on $\varepsilon$, but demanding high confidence is affordable, thanks to the logarithmic dependence on $\delta$. For example, in FPM, given $\varepsilon = 0.01$ and $\delta = 0.1$, $n \approx 14,979$, which means that for itemset $X$, if we sample 14,979 transactions from a data set, then its true support $\sigma(X)$ is beyond the range of $[\sigma(X) - 0.01, \sigma(X) + 0.01]$ with probability 0.1. In other words, $\sigma(X)$ is within $\pm\varepsilon$ of $\sigma(X)$ with high confidence 0.9.

Recent research has managed to reduce the disk I/O activity to two full scans over the database. Toivonen [Toi96] uses this statistical sampling technique in his algorithm. The algorithm makes only one full pass over the database. The idea is to pick a simple random sample, use it to determine all frequent patterns that probably hold in the whole database, and then to verify the results with the rest of the database in the second pass. The algorithm thus produces exact frequent patterns in one full pass over the database. In those rare cases where the sampling method does not produce all frequent patterns, the missing patterns can be found in a second pass. Toivonen's algorithm provides strong but probabilistic guarantees on the accuracy of its frequency counts.

In [ZPLO96], further study on the usefulness of sampling over large transactional data is carried out. They concluded that simple random sampling can reduce the I/O cost and computation time for association rule mining. This work is complementary to the approach in [Toi96], and can help in determining a better support value or sample size. The authors experimentally evaluate the effectiveness of sampling on different databases, and study the relationship between the performance, the accuracy and confidence of the chosen sample.

In [YCLZ04], Yu et al. proposed FPDM, which is perhaps the first false-negative oriented approach for mining frequent patterns over data streams. FPDM utilizes the Chernoff bound to achieve the quality control for finding frequent patterns. Their algorithm does not generate any false positive, and has a high-probability to find patterns which are truly frequent. In particular, they use a user-defined parameter $\delta$ to govern the probability to find the frequent patterns at support threshold $\sigma_{min}$. Specifically, the mining result does not include any patterns whose frequency is less than $\sigma_{min}$, and includes any pattern whose frequency exceeds $\sigma_{min}$ with probability of at least 1 - $\delta$. Similar to the LCA [MM02], FPDM partitioned the data stream into equal sized segments (batches). The batch size is given as $k \cdot n_0$ where $n_0$ is the required number of observations in order to achieve Chernoff bound with the user defined $\delta$. Note that $k$ is a parameter to control the batch size. For each batch, FPDM uses non-streaming FPM algorithm such as the Apriori algorithm to mine all the locally frequent patterns whose support is no less than $\sigma_{min} - \varepsilon$. The set of locally frequent patterns is then merged with those frequent patterns obtained so far from the stream. If the total number of patterns kept for the stream is larger than $c \cdot n$, where $c$ is an empirically determined float number, then all patterns whose support is less than ($\sigma_{min} - \varepsilon$) are pruned. FPDM outputs those patterns whose frequency is no less than $\sigma_{min} \times N$ where $N$ is the number of transactions received so far.

### 3.4   Reservoir Sampling

Described by Fan et al., the classic reservoir sampling algorithm is a sequential sampling algorithm over a finite population of records, with the population size unknown [FMR62]. Assuming that a sample of size $n$ is desired, one would commence by placing the first $n$ encountered into the putative sample (the reservoir). Each subsequent record is processed and considered in turn. The $k$-th record is accepted with probability $n/k$. If accepted, it displaces a randomly chosen record from the putative sample. Fan et al. showed that this will produce a simple random sample.

In practice, most stream sizes are unknown. Johnson et al. suggest that it is useful to keep a fixed-size sample [JMR05]. The issue of how to maintain a sample of a specified size over data that arrives online has been studied in the past. The standard solution is to use reservoir sampling of J.S. Vitter [Vit85]. The technique of reservoir sampling is, in one sequential pass, to select a random sample of $n$ transactions from a data set of $N$ transactions where $N$ is unknown and $n \ll N$. Vitter introduces Algo-Z, which allows random sampling of streaming data. The sample serves as a reservoir that buffers certain transactions from the data stream. New transactions appearing in the stream may be captured by the reservoir, whose limited capacity then forces an existing transaction to exit the reservoir. Variations on the algorithm allow it to idle for a period of time during which it only counts the number of transactions that have passed by. After a certain number of transactions are scanned, the algorithm can awaken to capture the next transaction from the stream. Reservoir Sampling can be very efficient, with time complexity less than linear in the size of the stream. Since 1985, there have been some modifications of Vitter's algorithm to make it more efficient. For example, in [Li94], Algo-Z, is modified to give a more efficient algorithm, Algo-K. Additionally, two new algorithms, Algo-L and Algo-M, are proposed. If the time for scanning the data set is ignored, all the four algorithms (K, L, M, and Z) have expected CPU time $O(n(1 + log(\frac{N}{n})))$, which is optimum up to a constant factor.

In [BDM02], "chain-sample" is proposed for a moving window of recent transactions from a data stream. If the $i^{th}$ transaction is selected for inclusion in the sample, another index from the range $i + 1, ..., i + n$ is selected which will replace in the event of its deletion. When the transaction with the selected index arrives, the algorithm stores it in memory and chooses the index of the transaction that will replace *it* when it expires and so on, thus building a chain of transactions to use in case of the expiration of the current transaction in the sample. The sample as a result will be a random sample. So, although this is able to overcome the problem of deletion in reservoir sampling, it still suffers from the problems that a random sample suffers from, namely, random fluctuations particularly when the sample size is small, and inability to handle noise. In [POSG04], a reservoir sampling with replacement for streaming data is proposed. It allows duplicates in the sample. The proposed method maintains a random sample with replacement at any given time. Although the difference between sampling with and without replacement is negligible with large population size, the authors argue that there are still situations when sampling with replacement is preferred.

In [Agg06], the author proposed a new method for biased reservoir based sampling of data streams. This technique is especially relevant for continual streams which gradually evolve over time, as a result of which the old data becomes obsolete for a variety of data mining applications. While biased reservoir sampling is a very difficult problem (with the one pass constraint), the author demonstrates that it is possible to design very efficient replacement algorithms for the important class of memory-less bias functions. In addition, the incorporation of bias results in upper bounds on reservoir sizes in many cases. In the special case of memory-less bias functions, the maximum space requirement is constant even for an infinitely long data stream. This is a nice property, since it allows for easy implementation in a variety of space-constrained scenarios.

### 3.5 Summary

The high complexity of the FPM problem handicaps the application of the stream mining techniques. We note that a review of existing techniques is necessary in order to research and develop efficient mining algorithms and data structures that are able to match the processing rate of the mining with high speed data streams. We have described the prior works related to our research. We have presented a number of the state-of-the-art algorithms on FPM and sampling over data streams. We summarize them in Table 1. Note that some of these general purpose sampling algorithms such as Algo-Z existed long before FPM become popular. However, they have proven to be useful tools for solving data stream problems.

## 4   Distance Based Sampling

Recent work in the field of approximate aggregation processing suggests that the true benefits of sampling might be well achieved when the sampling technique is tailored to the specific problem at hand [AGP00]. In this respect, we propose a distance based sampling that is designed to work "count" data set, that is, data set in which there is a base set of "items" and each data element (or transaction) is a vector of items. Note that in this paper, a sample refers to a set of transactions. Usually, for distance based sampling, the strategy is to produce a sample whose "distance" from the complete database is minimal. The main challenge is to find an appropriate distance function that can accurately capture the difference between the sample $(S)$ and all the transactions seen so far $(Ds)$ in the stream.

**Table 1.** Overview of FPM and sampling algorithms over data streams

| Authors | Algorithm Name | Window Model | Strategy |
|---|---|---|---|
| Manku and Motwani [MM02] | LCA | Landmark | Approximate Counting |
| Li et .al [LLS04] | DSM-FI | Landmark | Approximate Counting |
| Li et .al [LLS05] | DSM-MFI | Landmark | Approximate Counting |
| Manku and Motwani [MM02] | Sticky | Landmark | Sampling Related |
| Bronnimann et .al [BCD+03] | $EASE$ | Landmark | Sampling Related |
| Toivonen [Toi96] | not provided | Landmark | Sampling Related |
| Yu et al. [YCLZ04] | FPDM | Landmark | Sampling Related |
| Vitter [Vit85] | Algo-Z | Landmark | Sampling Related |
| Li [Li94] | Algo-K, Algo-L and Algo-M | Landmark | Sampling Related |
| Park et al. [POSG04] | RSWR | Landmark | Sampling Related |
| Chang and Lee [CL04] | not provided | Sliding | Approximate Counting |
| Chang and Lee [CL03a] | $estWin$ | Sliding | Approximate Counting |
| Chi et al. [CWYM04] | MOMENT | Sliding | Approximate Counting |
| Chang and Lee [CL03b] | $estDec$ | Damped | Approximate Counting |
| Aggarwal [Agg06] | not provided | Damped | Sampling Related |
| Giannella et al. [HPY00] | FP-Stream | Titled-time | Approximate Counting |

For distance based sampling, the basic intuition is that if the distance between the relative frequency of the items in $S$ and the corresponding relative frequency in $Ds$ is small, then $S$ is a good representative of $Ds$. Ideally, one needs to compare the frequency histograms of all possible itemsets: 1-itemsets, 2-itemsets, 3-itemsets, and so on. But experiments suggest that often it is enough to compare the 1-itemset histograms [BCD+03, DN06]. The proposed method, although based on 1-itemset histograms, can be easily extended to histograms for higher number of itemsets by simply adding them in constructing the histogram. On the flip side, histograms for higher itemsets are computationally expensive.

$$Dist_2(S, Ds) = \sum_{A \in \mathcal{I}} (Sup(A; S) - Sup(A; Ds))^2 \qquad (3)$$

Here, the relative frequency of item $A$ in $S$ and $Ds$ is given by $Sup(A; S) = \frac{freq(A;S)}{|S|}$ and $Sup(A; Ds) = \frac{freq(A;Ds)}{|Ds|}$, respectively. $freq(A; U)$ is the frequency of $A$ in a set of transactions $U$. The sum of squared distances variant ($Dist_2$) is one of the easiest way to measure the distance between vectors [CHS02]. Since we are aiming to minimize the distance between the item frequencies in the data stream and the sample, using $Dist_2$ is the natural way to measure the combined distance for all items. In this paper, we shall use Eq. 3 to measue the distant between $S$ and $Ds$ where $S$ is the working sample (to which the new transactions are being added or removed from), and $Ds$ is the reference set of transactions against which the accuracy of the working sample is compared.

## 4.1 Distance Based Sampling for Streaming Data

Like any other reservoir sampling method designed for data stream, the initial step of distance based sampling for streaming data (*DSS*) is to insert the first $n$ transactions into a "reservoir". The rest of the transactions are processed sequentially; transactions can be selected for the reservoir only as they are processed. Because $n$ is fixed, whenever there is an insertion of transaction, there is sure to be a deletion. In *DSS*, a local histogram ($Hist_L$) and a global histogram ($Hist_G$) are employed to keep track of the frequency of items generated by $S$ and $Ds$ respectively. Ideally, for a sample to be a good representation of the entire data, the discrepancy between $Hist_L$ and $Hist_G$ should be small. In other words, both the structure of $Hist_L$ and $Hist_G$ should look similar. Any insertion or deletion of transaction on the sample will affect the shape of $Hist_L$.

**Table 2.** Conceptual Ranking of Transactions

| Rank | Distance | *Tid* |
|------|----------|-------|
| $1^{st}$ | 25.00 | 10 |
| $2^{nd}$ | 15.00 | 20 |
| : | : | : |
| $n^{th}$ | 2.00 | 3 |

To maintain the sample, *DSS* prevents an incoming transaction from entering $S$ if its existence in $S$ increases the discrepancy. In addition, *DSS* helps to improve the quality of the sample by deleting transaction whose elimination from $S$ maximally reduces (or minimally increases) the discrepancy. Therefore, there is a ranking mechanism in *DSS* to rank the transactions in $S$ so that the "weakest" transaction can be replaced by the incoming transaction if the incoming transaction is better. The transactions in the initial sample are ranked by "leave-one-out" principle, i.e., distance is calculated by leaving out the transaction. Higher ranks are assigned to the transactions removal of which leads to higher distance. Mathematically, the distance is calculated as follows for ranking:

$$Dist_t = Dist_2\left(S - \{t\}, Ds\right) \tag{4}$$

where $t \in S$. For the initial ranking, both $S$ and $Ds$ contain the first $n$ transactions. Table 2 shows the conceptual idea of ranking. *Tid* 10 is ranked highest because its removal produces the maximum distance of 25.0. Similarly, *Tid* 3 is ranked lowest because its removal produces the minimum distance of 2.0. Let $LRT$ denotes the lowest ranked transaction.

When a new transaction $t_{new}$ arrives, $Hist_G$ is immediately updated. A decision is made whether to keep it in the sample by comparing the distances computed when $Hist_L$ is 'with' and 'without' $t_{new}$. First, we use 5 to calculate the distance between $Hist_G$ and $Hist_L$ when $t_{new}$ is absent. Next, $LRT$ is temporarily removed from the sample. We use 6 to calculate the distance between $Hist_G$ and $Hist_L$ when $t_{new}$ is present. Note that because of the incoming $t_{new}$, the $Ds$ in 5 and 6 is updated ($Ds = Ds + t_{new}$). If $Dist_{without\_t_{new}} > Dist_{with\_t_{new}}$, then $t_{new}$ is selected to replace $LRT$ in the current sample. $LRT$ is permanently removed. $t_{new}$ will be ranked in the sample using $Dist_{without\_t}$ value. On the other hand, if $Dist_{without\_t_{new}} \leq Dist_{with\_t_{new}}$, then $t_{new}$ is rejected and $LRT$ is retained in the reservoir.

$$Dist_{without\_t_{new}} = Dist_2(S, Ds) \tag{5}$$

$$Dist_{with\_t_{new}} = Dist_2((S - LRT + t_{new}), Ds) \tag{6}$$

Ideally, all transactions in the current sample should be re-ranked after each incoming transaction is processed because $Hist_G$ is modified. Re-ranking is done by recalculating the distances using 'leave-it-out' principle for all transactions in the current sample and those which are rejected already. But this can be computationally expensive because of the nature of data stream. We need a trade-off between accuracy and speed. Thus, re-ranking is done after selecting $\mathcal{R}$ new transactions in the sample. We do not consider the rejected transactions while counting $\mathcal{R}$. A good choice of $\mathcal{R}$ is 10. DSS tries to ensure that the relative frequency of every item in the sample is as close as possible to that of the original data stream. In the implementation, DSS stores the sample in an array of pointers to structures which hold the transactions and their distance values. Initial ranking and re-ranking of the sample according to the distances involves two steps. The first step is to calculate the distances of the transactions in the sample and the second step is to sort the sample by increasing distances using the standard quick sort technique. When $Dist_2$ is well implemented, the distance based sampling can have computational cost at most $O(|t_{max}|)$, where $|t_{max}|$ is the maximal length of the transaction vector (see Section 4.2). The complete DSS algorithm is summarized as follows:

1. Insert the first $n$ transactions into $S$.
2. Initialize $Hist_G$ and $Hist_L$ to keep track of the number of transactions containing each item $\mathcal{A}$ in $S$ and $Ds$.
3. Rank the initial $S$ by 'leave-it-out' method using 4.
4. Read the next incoming $t_{new}$.
5. Include $t_{new}$ into $Hist_G$.
6. Compare the distances of $S$ 'without' and 'with' $t_{new}$ using 5 and 6.
   a. If $Dist_{without\_t_{new}} > Dist_{with\_t_{new}}$, then replace LRT with $t_{new}$ and update $Hist_L$.
   b. Else, reject $t_{new}$.
7. If $\mathcal{R}$ new transactions are already selected, re-rank the transactions in the current sample. Eg. $\mathcal{R} = 10$.
8. Repeat steps 4 to 7 for every subsequent incoming transaction $t_{new}$.

## 4.2  Complexity Analysis

When finding $LRT$, we need to compute $Dist_2$ for each $t \in S$. This is the distance between $Hist_G$ and $Hist_L$ when a transaction $t$ is removed from $S$. Here, $S_t$ denotes $S - \{t\}$. We let $\mathcal{F}_{AS}$, $\mathcal{F}_{AS_t}$ and $\mathcal{F}_{ADs}$ represent the absolute frequency of item $\mathcal{A}$ in $S$, $S_t$ and $Ds$ respectively. Note that,

$$\mathcal{F}_{AS_t} = \begin{cases} \mathcal{F}_{AS} - 1 & \text{if } \mathcal{A} \in t \\ \mathcal{F}_{AS} & \text{else } \mathcal{A} \notin t. \end{cases}$$

To search for the weakest transaction in $S$, the set of $n$ distances that are generated from Eq 4 has to be compared with one another such that

$$t^* = \underset{1 \le t \le n}{\operatorname{argmin}} \; Dist_2(S - t, Ds) = \underset{1 \le t \le n}{\operatorname{argmin}} \sum_{A \in \mathcal{I}} \left( \frac{\mathcal{F}_{AS_t}}{|S_t|} - \frac{\mathcal{F}_{ADs}}{|Ds|} \right)^2. \tag{7}$$

Unfortunately, the determination of $t^*$ can be computationally costly. The worst case time complexity is $O(n.|\mathcal{I}|)$, where $|\mathcal{I}| \gg 1$. However, we note that even though removing a transaction from $S$ will affect relative frequencies of all items, most absolute frequencies will remain unchanged. Only those items contained within the transaction $t$ are affected. In addition, we can make use of the fact that, in general,

$$\underset{x \in U}{\operatorname{argmin}} f(x) = \underset{x \in U}{\operatorname{argmin}} cf(x) + d. \tag{8}$$

For any constant $c$ and real number $d$ that we introduced, the final outcome will still remain the same. With this understanding, we can rewrite 7 as

$$t^* = \underset{1 \le t \le n}{\operatorname{argmin}} \sum_{A \in \mathcal{I}} \left( \left( \mathcal{F}_{AS_t} - \frac{\mathcal{F}_{ADs}}{|Ds|} |S_t| \right)^2 - \left( \mathcal{F}_{AS} - \frac{\mathcal{F}_{ADs}}{|Ds|} |S_t| \right)^2 \right)$$

$$t^* = \underset{1 \le t \le n}{\operatorname{argmin}} \sum_{A \in t} \left( 1 - 2\mathcal{F}_{AS} + 2\frac{\mathcal{F}_{ADs}}{|Ds|} |S_t| \right)$$

$$\tag{9}$$

Clearly, from the above representation of $t^*$, it is possible to reduce the worst-case cost of ranking from $O(n.|\mathcal{I}|)$ to $O(n.|T_{max}|)$. Similarly, for comparing between $LRT$ and $t_{new}$, we can apply the same strategy. The cost will then be $O(|T_{max}|)$.

### 4.3   Handling Noise in *DSS*

Noise here means a random error or variance in a measured variable [HK06a]. The data object that holds such noise is a noisy data object. Such noisy transactions have very little or no similarity with other transactions. This "aloof" nature of noisy transactions is used to detect and remove them. Noise detection and removal has been studied abundantly, for example in [KM03, ZWKS07]. These methods typically employ a similarity measurement and a similarity threshold to determine whether an object is noise, that is if a data object has lesser similarity than the threshold, it is considered noise.

In this paper the focus is to maintain a small sample of transactions over a streaming data. *DSS* not only maintains a good representative sample, it also removes noise by not selecting them in the sample. *DSS* banks on the *aloofness* of the transactions to determine whether an incoming transaction is corrupted by noise. Removal of a noisy transaction will induce smaller distance than removal of a *bona fide* transaction. As *DSS* selects transactions into the reservoir sample by the distance it induces, it naturally rejects the noisy transactions. Later experimental results show that *DSS* has a better ability to handle noise than LCA and Algo-Z.

## 5   Performance Study

This section describes the experimental comparison between *DSS* and LCA in the context of FPM. In addition, we also compared *DSS* with simple random sampling (SRS) using Algo-Z. All experiments were performed on a 1.7GHz CPU Dell PC with 1GB of main memory and running on the Windows XP platform. We used the code from the IBM QUEST project [AS94] to generate the datasets. We set $\mathcal{I} = 10k$ unique items. The two datasets that we generated are $T10I3D2000K$ and $T15I7D2000K$. Note that $T15I7D2000K$ is a much denser dataset than $T10I3D2000K$ and therefore requires more time to process. In addition, to verify the performance of all the 3 algorithms on real world dataset, we used a coded log of a clickstream data (denoted by Kosarak) from a Hungarian on-line news portal [Bod03]. This database contains 990002 transactions with average size 8.1. We fixed the minimum support threshold $\sigma_{min} = 0.1\%$ and the batch size $B_{size} = 200K$. For comparison with LCA, we set $\epsilon = 0.1\sigma_{min}$ which is a popular choice for LCA.

### 5.1   Performance Measure

To measure the quality of the mining results, we use two metrics: the recall and the precision. Given a set $AFI_{true}$ of all true frequent itemsets and a set $AFI_{appro}$ of all approximate frequent itemsets obtained in the output by some mining algorithms, the recall is $\frac{|AFI_{true} \cap AFI_{appro}|}{|AFI_{true}|}$ and the precision is $\frac{|AFI_{true} \cap AFI_{appro}|}{|AFI_{appro}|}$. If the recall equals 1, the results returned by the algorithm contains all true results. This means no false negative. If the precision equals 1, all the results returned by the algorithm are some or all of the true results, i.e., no false positive is generated. To evaluate the quality of the sample, we can apply the symmetric difference

$$SD = \frac{|(AFI_{true} - AFI_{appro}) \cup (AFI_{appro} - AFI_{true})|}{|AFI_{true}| + |AFI_{appro}|}. \tag{10}$$

Note that the above expression is nothing more than the ratio of the sum of missed and false itemsets to the total number of $AFI$ [BCD+03, LCK98]. The $SD$ is zero when $AFI_{true} = AFI_{appro}$ and the $SD$ is 1 when $AFI_{true}$ and $AFI_{appro}$ are disjoint. The greater the value of $SD$, the greater is the dissimilarity between $AFI_{true}$ and $AFI_{appro}$ (and vice versa). Alternately, we can define the overall accuracy with

$$Acc = 1 - SD. \tag{11}$$

### 5.2   Time Measurements

We report the experimental results for the three data sets described previously. Figure 5 depicts the number of frequent patterns uncovered during the operation of the three algorithms for the three different data sets. The set of frequent patterns are broken down according to the pattern length. Note that for *DSS* and Algo-Z, the size of the reservoir is 5k transactions. In the figure, ORG indicates the true size of frequent itemsets generated when the Apriori algorithm operates on the entire data set. Interestingly, we can observe that the general trends of the sampled data sets resemble the true result. This is also similar with the output from LCA. However, Algo-Z tends to drift very far away from

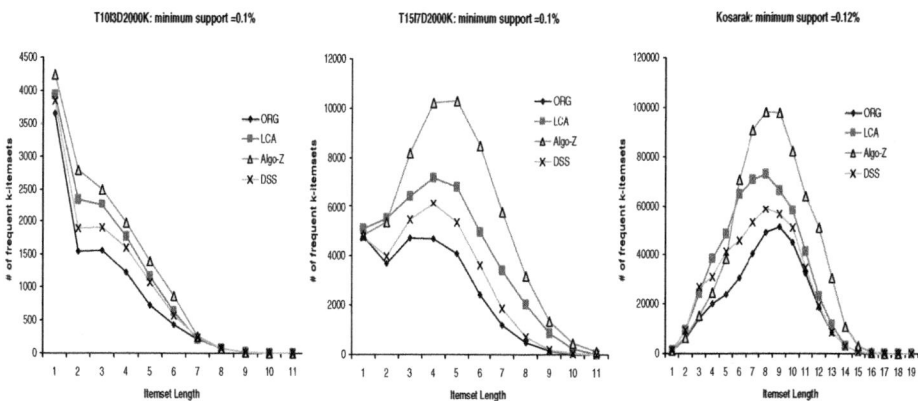

**Fig. 5.** Itemset size vs Number of Frequent Itemsets

ORG indicating more false positives are generated. LCA is caught in between the two sampling algorithms. *DSS* gives the most appealing results in terms of fidelity. The graphs from *DSS* are the closest to ORG.

Figure 6 shows the execution time on the three data sets. The results of the algorithms are computed as an average of 20 runs. Each run corresponds to a different shuffle of the input data set. Note that for *DSS* and Algo-Z, the execution time consists of the time spent for obtaining the sample as well as using the Apriori algorithm to generate the frequent patterns ($AFI_{appro}$) from the sample. As we can see, the cumulative execution times of *DSS*, Algo-Z and LCA grow linearly with the number of transactions processed in the streams. In particular, Algo-Z is the fastest because it only needs to maintain a sample by randomly selecting transactions to be deleted or inserted in the reservoir. There is no real processing work to be done on any incoming transaction. Unlike Algo-Z, *DSS* processes every incoming transaction by computing the distance it may cause if included in the sample. As a result, its processing speed is slower than Algo-Z. However, the slowest algorithm is LCA. In all three data sets, LCA spent the most amount of time.

### 5.3 Accuracy Measurements

Figure 7 displays the accuracies of the three algorithms against the number of transactions processed. Here, *Acc* is computed using 11. As expected, Algo-Z achieved the worst performance. As the number of transactions to be processed increases, we see that its accuracy drops significantly. Since the reservoir is a fixed size, the sampling ratio decreases at every test point. From the graph, *DSS* achieved good accuracy even for small sampling ratios. Its accuracy remains almost stable for all the test points. Although LCA guarantees 100% recall, its performance was heavily affected by its poor precision. This can be clearly seen in the figure. From Figure 7, it is clear that the accuracy of LCA was dragged down due to its low precision value. For all the three data sets, its performance is lower than *DSS*.

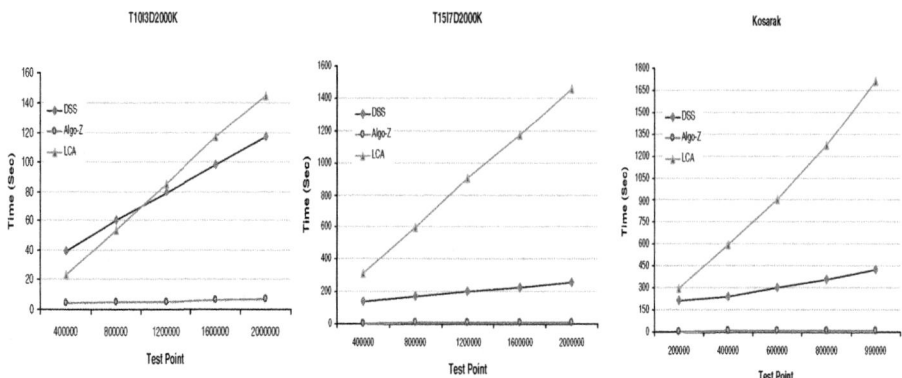

**Fig. 6.** Execution time on $T10I3D2000K$, $T15I7D2000K$ and Kosarak

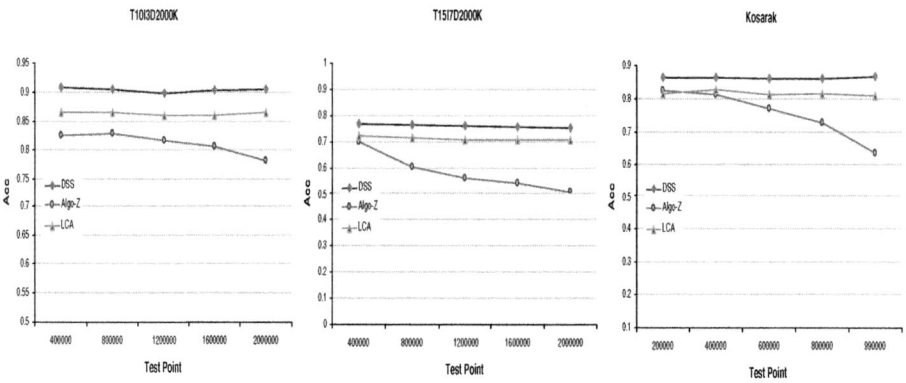

**Fig. 7.** Accuracy on $T10I3D2000K$, $T15I7D2000K$ and Kosarak

## 5.4   Varying $\mathcal{R}$

When running $DSS$, there is always a need to re-calculate the distance of the transactions in the sample because the global histogram will be slightly modified whenever a new transaction arrives. Unfortunately, to re-rank every transaction in the sample for every incoming transaction will be very costly. To cope with this problem, the parameter, $\mathcal{R}$, is introduced. The idea is very simple. If $\mathcal{R}$ new transactions have been inserted into the sample, we will activate the ranking mechanism. $\mathcal{R}$ is the only parameter in $DSS$. Figure 8 illustrates the impact of varying the value of $\mathcal{R}$ in $DSS$. We provide four variations (1, 5, 10 and 15) of $\mathcal{R}$ for the data set $T15I7D2000K$. From the Figure 8a, we see an obvious gap for $\mathcal{R} = 10$ and $\mathcal{R} = 15$. However, the different in accuracy for $\mathcal{R} = 1$, $\mathcal{R} = 5$ and $\mathcal{R} = 10$ is not so apparent. In addition, as $\mathcal{R}$ gets smaller, the execution time increases. This suggests that the value 10 is a suitable default value for $\mathcal{R}$. Unlike LCA which has a drastic effect either on its speed or memory if $\epsilon$ is wrongly set, $\mathcal{R}$ is less sensitive.

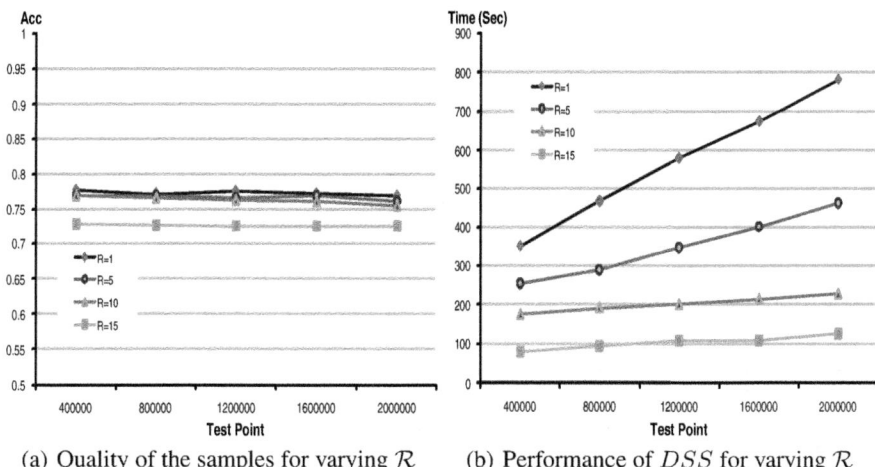

(a) Quality of the samples for varying $\mathcal{R}$    (b) Performance of $DSS$ for varying $\mathcal{R}$

**Fig. 8.** Impact of varying the value of $\mathcal{R}$

## 5.5  *DSS* with Higher Itemsets

So far, we have been discussing using the frequency of single itemsets to maintain the histogram. What about the support for 2-itemsets or k-itemsets? Logically, *DSS* having higher itemsets would perform better than one that is using only 1-itemsets. However, by considering all large itemsets in the histogram would be to generate all the $2^{\mathcal{I}}$ subsets of the universe of $\mathcal{I}$ items. It is not hard to observe that this approach exhibit complexity exponential in $\mathcal{I}$, and is quite impractical. For example, in our experiment, we have $\mathcal{I}$ = 10000 items. If we consider having 2-itemsets in the histogram, we would generate about 50 million of possible combinations. Note that it is still technically possible to maintain an array of 50 million 2-itemsets in a system. However, to go up a higher order will be impossible. In this section, we shall compare the performance of *DSS* with and without 2-itemsets in the histogram.

To maintain 50 million counters in $Hist_G$ and $Hist_L$, we can still adopt the trimming method discussed in Section 4.2. The extra work added to *DSS* is the generation of all possible 2-itemset combinations for all incoming transactions. For example, a transaction having item $A, B, C$ and $D$ will generate $AB, AC, AD, BC, BD$ and $CD$. In Figure 9, we compare the performance of *DSS* having 2-itemsets with one that uses only 1-itemsets on $T15I7D2000K$. From Figure 9a we see that *DSS* having 2-itemsets on average has a small increase of 2% in accuracy over normal *DSS*. In Figure 9b, we discover the price we need to pay for higher accuracy. The execution time for *DSS* having 2-itemsets increases tremendously. It took more than 2 hours to complete the sampling process where as a normal *DSS* finished in less than 5 minutes. Therefore, this explains why at our current research level we only focus on single itemsets.

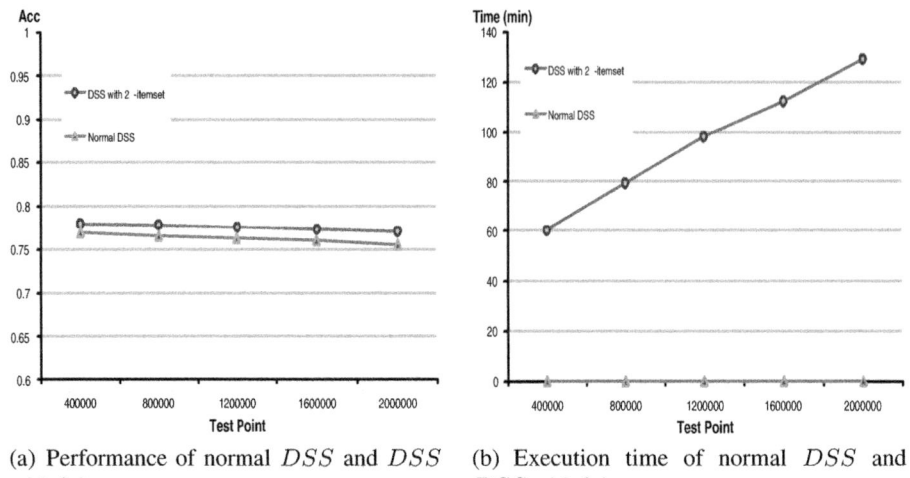

(a) Performance of normal $DSS$ and $DSS$ with 2-itemset

(b) Execution time of normal $DSS$ and $DSS$ with 2-itemset

**Fig. 9.** Normal $DSS$ vs. $DSS$ with 2-itemset

## 5.6   Handling Noise

Removing transactions that are corrupted with noise is an important goal of data cleaning as noise hinders most types of data analysis. This section shows how $DSS$ is not only able to produce high quality sample from normal data sets, but it is also able to cope with data sets having corrupted transactions by preventing these transactions from being inserted into the reservoir. To demonstrate the robustness of $DSS$ against noise, we let the three algorithms operate on a noisy data set. For this experiment, we added 5% of noise to $T15I7D2000K$. Noise was added using the $rand$ function. With a probability of 5%, we corrupt an item in a transaction by replacing it with any item in $\mathcal{I}$. Similar to the previous experiments, we made use of the true frequent patterns uncovered from the original data set to compare with the approximate frequent patterns uncovered from the corrupted data set. Figure 10 illustrates the performances of the three algorithms against noise. For reference, the results for the algorithms operating on the noise free data set are also included in the plot. From the graph, $DSS$ suffers the least in terms of accuracy when noise was added to the data. Its overall accuracy is maintained at about 75% and its maximun drop in performance is at most 2% when the test point is at 400k. However, for LCA and Algo-Z, the gap between the original result without noise and the result with noise is wider when compared with the one by $DSS$. The greatest drop in accuracy is by 5% for LCA when the test point is at 2000k and by 6% for Algo-Z when the test point is at 1600k.

To make the comparison complete, we introduced 5%, 10% and 20% of noise to the three data sets. Figures 11, 12 and 13, show the results by varying the noise level. In all the data sets, the performances of the three algorithms are affected by noise. Their performances declines as the noise level increases. However, a closer look at the results reveals that even when the noise level is set at 20%, the performance of $DSS$ does not suffer too badly as compared with Algo-Z and LCA. For example in Figure 13, when

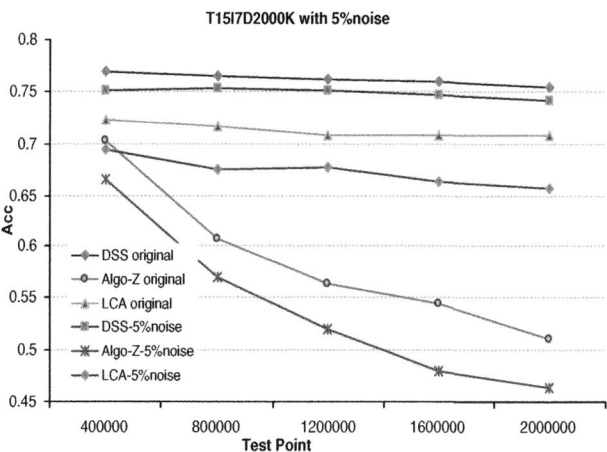

**Fig. 10.** Accuracy on corrupted data set

processing Kosarak data set having 20% of noise level, the drop in accuracy for *DSS* is at most by 5% (test point = 400k). However, Algo-Z suffers 21% loss in accuracy (test point = 990k) and LCA suffers 11% (test point = 400k).

In these experiments, we can see that *DSS* outperforms the other two algorithms in terms of noise resistance. We reason this is due to the ranking process of *DSS*. As *DSS* is a deterministic sampling, every incoming transaction is processed and ranked according to its importance in the sample. When an incoming transaction is corrupted, its presence in the reservoir would create a larger discrepancy between $Hist_L$ and $Hist_G$ than when it is not corrupted and thus its chance to be inserted into the reservoir is low. In other words, employing distance based sampling is beneficial as it helps in filtering out noisy data from the database. As for Algo-Z, it uses simple random sampling and therefore it is blind towards noisy data. This explains why its performance is the most unreliable. Similarly for LCA, data is processed in batches. In a single batch of transactions, there can be a mixture of actual transactions as well as some corrupted transactions. LCA only maintains a set of entries to keep track of the counts of those itemsets that it regards as significant patterns. However, there is no filtering process to distinguish between actual and corrupted transactions. Therefore its performance is poor particularly when a lot of noise is introduced.

### 5.7  Comparison with Theoretical Bounds

Zaki et al. [ZPLO96] set the sample size using Chernoff bounds and find that sampling can speed up mining of association rules. However, the size bounds are found too conservative in practice. Moreover, Chernoff bound assumes data independence. In reality, data in a data stream is most probably dependent or correlated. When data is dependent in a transactional data stream, the quality of the sample cannot be guaranteed. In the last part of this experimental evaluation, we shall compare the performance of distance based sampling with the theoretical bounds.

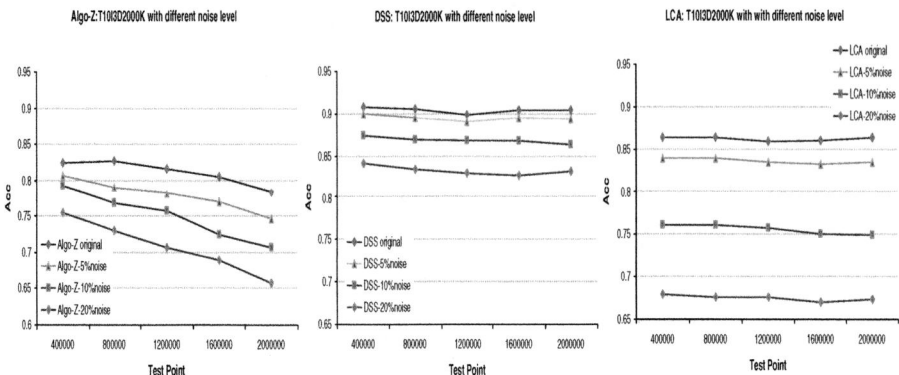

**Fig. 11.** Accuracy on $T10I3D2000K$ with different noise level

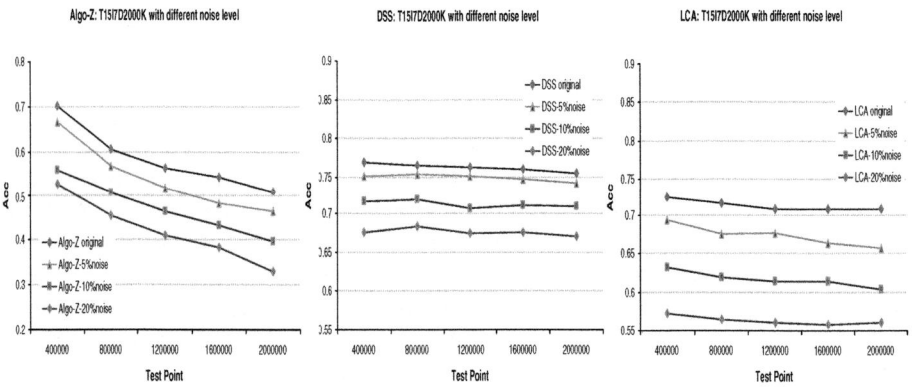

**Fig. 12.** Accuracy on $T15I7D2000K$ with different noise level

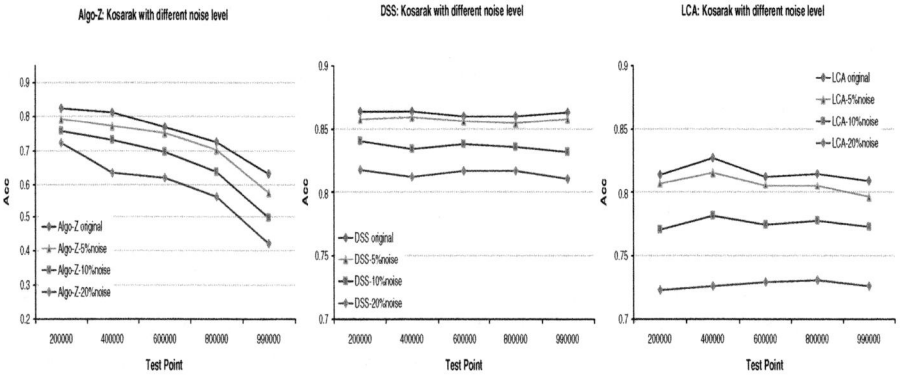

**Fig. 13.** Accuracy on Kosarak with different noise level

Denote by $\mathbf{X}$ the number of transactions in the sample containing the itemset $I$. Random variable $\mathbf{X}$ has a binomial distribution of $n$ trials with the probability of success $\sigma_{min}$. For any positive constant, $0 \leq \varepsilon \leq 1$, the Chernoff bounds [HR90] state that

$$P(\mathbf{X} \leq (1 - \varepsilon)n\sigma_{min}) \leq e^{-\varepsilon^2 n\sigma_{min}/2} \tag{12}$$

$$P(\mathbf{X} \geq (1 + \varepsilon)n\sigma_{min}) \leq e^{-\varepsilon^2 n\sigma_{min}/3} \tag{13}$$

Chernoff bounds provide information on how close the actual occurrence of an itemset in the sample is, compared to the expected count in the sample. *Accuracy* is given as $1 - \varepsilon$. The bounds also tell us the probability that a sample of size $n$ will have a given accuracy. We call this aspect *confidence* of the sample (defined as 1 minus the expression on the right hand side of the equations). Chernoff bounds give us two sets of confidence values. The first equation gives the lower bound – the probability that the itemset occurs less often than expected and the second one gives the upper bound – the probability that the itemset occurs more often than expected.

The following plots in Figure 14 show the results of comparing theoretical Chernoff bound with experimentally observed results. We show that for the databases we have considered the Chernoff bound is very conservative compared to the two sampling algorithms. Furthermore, we show that *DSS* samples are more accurate than reservoir sampling Z algorithm. We can obtain the theoretical confidence value by simply evaluating the right hand side of the equations. For example, for the upper bound the confidence $C = 1 - e^{-\varepsilon^2 n\sigma_{min}/3}$. We can obtain experimental confidence values as follows. We take $s$ samples of size $n$, and for each item we compute the confidence by evaluating the left hand side of the two equations as follows. Let $i$ denote the sample number, $1 \leq i \leq s$. Let $l_I(i) = 1$ if $(n\sigma_{min} - \mathbf{X}) \geq n\sigma_{min}\varepsilon$ in sample $i$, otherwise 0. Let $h_I(i) = 1$ if $(\mathbf{X} - n\sigma_{min}) \geq n\sigma_{min}\varepsilon$ in sample $i$, otherwise 0. The confidence can then be calculated as $1 - \sum_{i=1}^{m} h_I(i)/s$, for the upper bound. For our experiment we take $s = 100$ samples for both algorithms Z and *DSS* for each of the three data sets. We cover our discussion on all 1-itemsets. Using the theoretical and experimental approaches we determine the probabilities (1-confidence) and plot them in the following figures.

Figure 14 compares the distribution of experimental confidence of simple random sampling and *DSS* to the one obtained by Chernoff upper bounds. The graphs show the results using $T15I7D2000K$, $n = 2000$ with $\varepsilon = 0.01$. From the figure, Chernoff bounds, with a mean probability of 99.95%, suggests that this sample size is 'likely' unable to achieve the given accuracy. Obviously, this is very pessimistic and over conservative. In actual case, simple random sampling (or Algo-Z) and *DSS* gave a mean probability of 75% and 43% respectively.

Figures 15, 16 and 17, provide a broader picture of the large discrepancy between Chernoff bounds and experimental results. For the three data sets, we plot the mean of the probability distribution for different Epsilon ($\varepsilon$). Different values of sample size are used (from 0.1% to 10%). The higher the probability, the more conservative the approach is. So we can see that *DSS* samples are the most accurate, followed by Algo-Z samples, and the theoretical bounds are the most conservative.

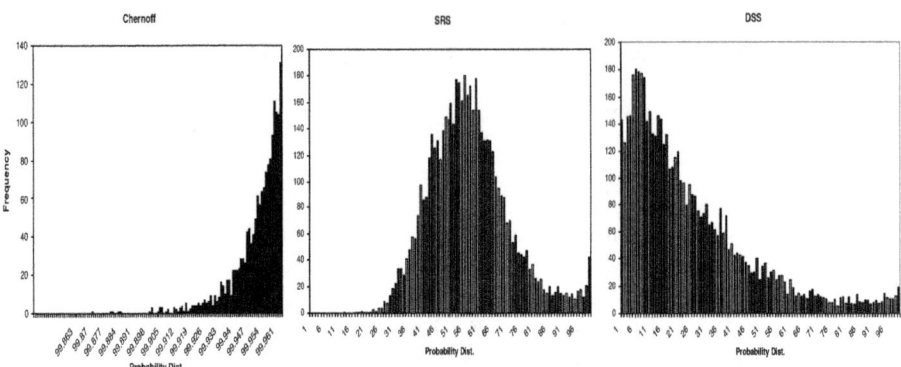

**Fig. 14.** Probability Distribution

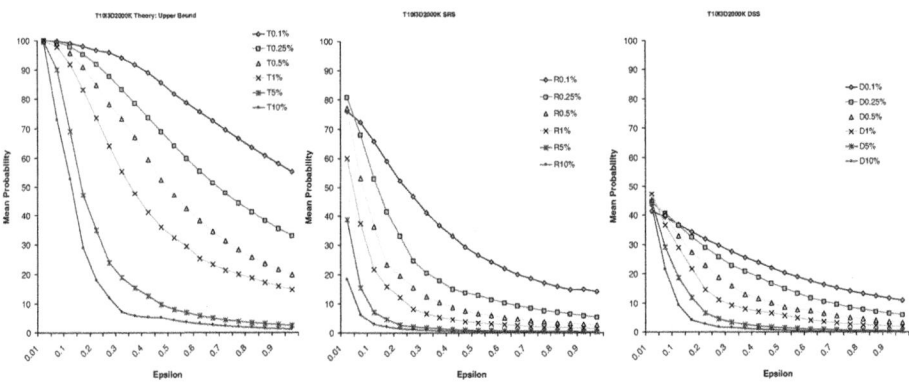

**Fig. 15.** $T10I3D2000K$: Epsilon vs. mean Probability

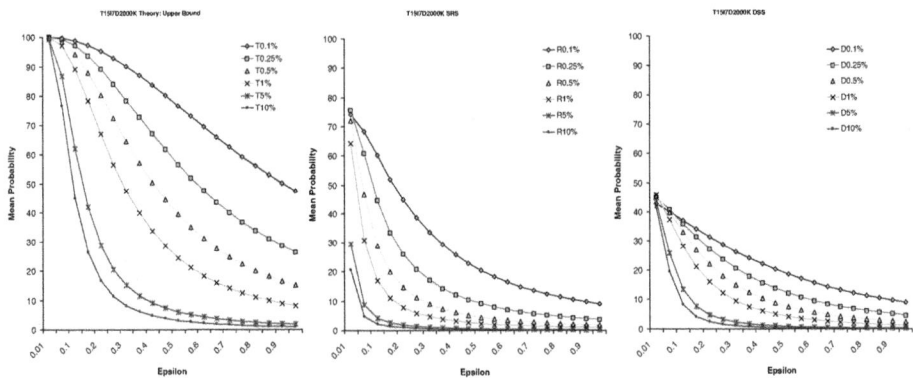

**Fig. 16.** $T15I7D2000K$: Epsilon vs. mean Probability

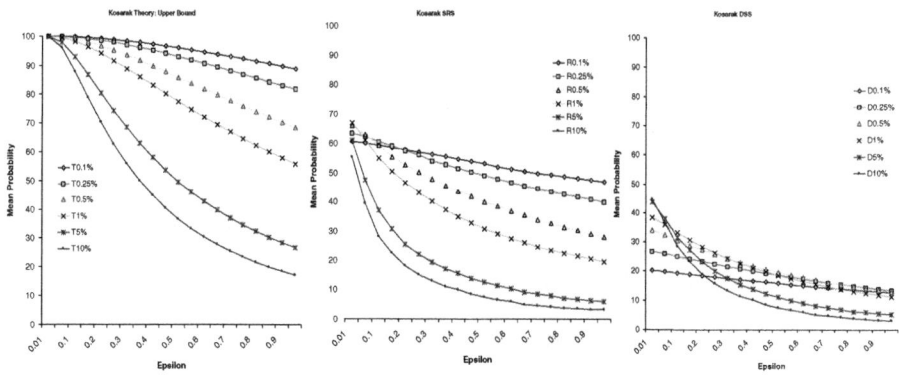

**Fig. 17.** Kosarak: Epsilon vs. mean Probability

# 6 Conclusions

There is a pressing need to have algorithms that handle data in an online fashion with a single scan of data stream whilst operating under resource constraints. In this paper, we focus on the problem of online mining for frequent patterns. Two prominent approaches have been discussed (approximate counting and sampling). We have compared 3 algorithms – LCA, Algo-Z and *DSS*. Moreover, we addressed the merits and the limitations and presented an overall analysis of these algorithms.

For approximate counting, we note that defining a proper value of $\epsilon$ is non-trivial. Although LCA is able to guarantee 100% recall, a slight variation on $\epsilon$ might result in a profound impact on its speed and precision. For sampling, random (Algo-Z) and deterministic (*DSS*) sampling are covered. Sampling is attractive because it permits normal mining algorithms that deal with static data to handle dynamic data (stream). If the sample size is small, it is advisable to adopt deterministic sampling than random sampling although deterministic sampling is more time consuming. An empirical study using both real and synthetic databases supports our claims of efficiency and accuracy. Random sampling produces sample of poor quality when the sample is small.

Results show that *DSS* is significantly and consistently more accurate than both LCA and Algo-Z, whereas LCA performs consistently better than Algo-Z. The sample generated by *DSS* is well suited for dataset having vector of items. However, beside data mining, there are other problems that are highly relevant for data streams. For example, we can explore distance based sampling to search for unusual data. This has application to network intrusion detection and can be considered for future extension of *DSS*.

# References

[Agg06]    Aggarwal, C.C.: On biased reservoir sampling in the presence of stream evolution. In: Proceedings of the 32nd International Conference on Very Large Data Bases, pp. 607–618 (2006)

[Agg07]    Aggarwal, C.C.: Data Streams: Models and Algorithms. Springer, Heidelberg (2007)

[AGP00]      Acharya, S., Gibbons, P.B., Poosala, V.: Congressional samples for approximate
             answering of group-by queries. In: Proceedings of the 2000 ACM SIGMOD Inter-
             national Conference on Management of Data, pp. 487–498 (2000)
[AS94]       Agrawal, R., Srikant, R.: Fast algorithms for mining association rules. In: Pro-
             ceedings of 20th International Conference on Very Large Data Bases, pp. 487–499
             (1994)
[AY06]       Aggarwal, C.C., Yu, P.: A survey of synopsis construction in data streams. Data
             Streams: Models and Algorithms, 169–208 (2006)
[BCD⁺03]     Bronnimann, H., Chen, B., Dash, M., Haas, P., Scheuermann, P.: Efficient data
             reduction with ease. In: Proceedings of the Ninth ACM SIGKDD International
             Conference in Knowledge Discovery and Data Mining, pp. 59–68 (2003)
[BDF⁺97]     Barbar'a, D., Dumouchel, W., Faloutsos, C., Haas, P.J., Hellerstein, J.M., Ioannidis,
             Y., Jagadish, H.V., Johnson, T., Ng, R., Poosala, V., Ross, K.A., Sevcik, K.C.: The
             new jersey data reduction report. IEEE Data Engineering Bulletin 20, 3–45 (1997)
[BDM02]      Babcock, B., Datar, M., Motwani, R.: Sampling from a moving window over
             streaming data. In: Proceedings of 13th Annual ACM-SIAM Symposium on Dis-
             crete Algorithms (2002)
[Bod03]      Bodon, F.: A fast apriori implementation. In: Proceedings of the IEEE ICDM
             Workshop on Frequent Itemset Mining Implementations, FIMI'03 (2003)
[CDG07]      Calders, T., Dexters, N., Goethals, B.: Mining frequent itemsets in a stream. In:
             IEEE International Conference on Data Mining (ICDM'07), pp. 83–92 (2007)
[CDH⁺02]     Chen, Y., Dong, G., Han, J., Wah, B.W., Wang, J.: Multi-dimensional regression
             analysis of time-series data streams. In: Proceedings of 28th International Confer-
             ence on Very Large Data Bases, pp. 323–334 (2002)
[CHS02]      Chen, B., Haas, P.J., Scheuermann, P.: A new two-phase sampling based algorithm
             for discovering association rules. In: Proceedings of the Eighth ACM SIGKDD
             International Conference on Knowledge Discovery and Data Mining, pp. 462–468
             (2002)
[CKN06]      Cheng, J., Ke, Y., Ng, W.: Maintaining frequent itemsets over high-speed data
             streams. In: Proceedings of the 10th Pacific-Asia Conference on Knowledge Dis-
             covery and Data Mining, pp. 462–467 (2006)
[CKN08]      Cheng, J., Ke, Y., Ng, W.: A survey on algorithms for mining frequent itemsets
             over data streams. An International Journal of Knowledge and Information Systems
             (2008)
[CL03a]      Chang, J.H., Lee, W.S.: *estWin*: adaptively monitoring the recent change of fre-
             quent itemsets over online data streams. In: CIKM, pp. 536–539 (2003)
[CL03b]      Chang, J.H., Lee, W.S.: Finding recent frequent itemsets adaptively over online
             data streams. In: Proceedings of the Seventh ACM SIGKDD International Confer-
             ence on Knowledge Discovery and Data Mining, pp. 487–492 (2003)
[CL04]       Chang, J.H., Lee, W.S.: A sliding window method for finding recently frequent
             itemsets over online data streams. Journal of Information Science and Engineeer-
             ing 20(4), 753–762 (2004)
[CS03]       Cohen, E., Strauss, M.: Maintaining time-decaying stream aggregates. In: Pro-
             ceedings of the Twenty-Second ACM SIGACT-SIGMOD-SIGART Symposium on
             Principles of Database Systems, pp. 223–233 (2003)
[CWYM04]     Chi, Y., Wang, H., Yu, P.S., Muntz, R.R.: Moment: Maintaining closed frequent
             itemsets over a stream sliding window. In: Proceedings of the 4th IEEE Interna-
             tional Conference on Data Mining (ICDM 2004), pp. 59–66 (2004)
[DH00]       Domingos, P., Hulten, G.: Mining high-speed data streams. In: Proceedings of
             ACM SIGKDD International Conference in Knowledge Discovery and Data Min-
             ing, pp. 71–80 (2000)

[DN06]      Dash, M., Ng, W.: Efficient reservoir sampling for transactional data streams. In: IEEE ICDM workshop on Mining Evolving and Streaming Data, pp. 662–666 (2006)

[FMR62]     Fan, C.T., Muller, M.E., Rezucha, I.: Development of sampling plans by using sequential (item by item) selection techniques and digital computers. Journal of the American Statistical Association (1962)

[GHP$^+$03]   Giannella, C., Han, J., Pei, J., Yan, X., Yu, P.S.: Mining frequent patterns in data streams at multiple time granularities. In: Next Generation Data Mining. AAAI/MIT (2003)

[HCXY07]    Han, J., Cheng, H., Xin, D., Yan, X.: Frequent pattern mining: current status and future directions. Data Min. Knowl. Discov. 15(1), 55–86 (2007)

[HK06a]     Han, J., Kamber, M.: Data Mining: Concepts and Techniques, 2nd edn. Morgan Kaufmann Publishers, San Francisco (2006)

[HK06b]     Hwang, W., Kim, D.: Improved association rule mining by modified trimming. In: Proceedings of Sixth IEEE International Conference on Computer and Information Technology, CIT (2006)

[HPY00]     Han, J., Pei, J., Yin, Y.: Mining frequent patterns without candidate generation. In: 2000 ACM SIGMOD Intl. Conference on Management of Data, pp. 1–12 (2000)

[HR90]      Hagerup, T., Rub, C.: A guided tour of chernoff bounds. Information Processing Letters, 305–308 (1990)

[JG06]      Jiang, N., Gruenwald, L.: Research issues in data stream association rule mining. SIGMOD Record 35, 14–19 (2006)

[JMR05]     Johnson, T., Muthukrishnan, S., Rozenbaum, I.: Sampling algorithms in a stream operator. In: Proceedings of the 2005 ACM SIGMOD International Conference on Management of Data, pp. 1–12 (2005)

[KK06]      Kotsiantis, S., Kanellopoulos, D.: Association rules mining: A recent overview. GESTS International Transactions on Computer Science and Engineering 32(1), 71–82 (2006)

[KM03]      Kubica, J.M., Moore, A.: Probabilistic noise identification and data cleaning. In: Proceedings of International Conference on Data Mining (ICDM), pp. 131–138 (2003)

[KSP03]     Karp, R.M., Shenker, S., Papadimitriou, C.H.: A simple algorithm for finding frequent elements in streams and bags. ACM Transactions on Database Systems 28(1), 51–55 (2003)

[LCK98]     Lee, S.D., Cheung, D.W.-L., Kao, B.: Is sampling useful in data mining? a case in the maintenance of discovered association rules. Data Mining and Knowledge Discovery 2(3), 233–262 (1998)

[Li94]      Li, K.-H.: Reservoir sampling algorithms of time complexity o(n(1 + log(n/n))). ACM Transactions on Mathematical Software 20(4), 481–493 (1994)

[LLS04]     Li, H.F., Lee, S.Y., Shan, M.K.: An efficient algorithm for mining frequent itemsets over the entire history of data streams. In: Proc. of First International Workshop on Knowledge Discovery in Data Streams (2004)

[LLS05]     Li, H.F., Lee, S.Y., Shan, M.K.: Online mining (recently) maximal frequent itemsets over data streams. In: RIDE, pp. 11–18 (2005)

[MG82]      Misra, J., Gries, D.: Finding repeated elements. Scientific Computing Programming 2(2), 143–152 (1982)

[MM02]      Manku, G.S., Motwani, R.: Approximate frequency counts over data streams. In: Proceedings of 28th International Conference on Very Large Data Bases, pp. 346–357 (2002)

[MTV94]    Mannila, H., Toivonen, H., Inkeri Verkamo, A.: Efficient algorithms for discovering association rules. In: Fayyad, U.M., Uthurusamy, R. (eds.) AAAI Workshop on Knowledge Discovery in Databases (KDD-94), pp. 181–192 (1994)

[ND06]     Ng, W., Dash, M.: An evaluation of progressive sampling for imbalanced data sets. In: IEEE ICDM Workshop on Mining Evolving and Streaming Data, pp. 657–661 (2006)

[ND08]     Ng, W., Dash, M.: Efficient approximate mining of frequent patterns over transactional data streams. In: Song, I.-Y., Eder, J., Nguyen, T.M. (eds.) DaWaK 2008. LNCS, vol. 5182, pp. 241–250. Springer, Heidelberg (2008)

[OR95]     Olken, F., Rotem, D.: Random sampling from databases - a survey. Statistics and Computing 5, 25–42 (1995)

[Par02]    Parthasarathy, S.: Efficient progressive sampling for association rules. In: IEEE International Conference on Data Mining (ICDM'02), pp. 354–361 (2002)

[PJO99]    Provost, F.J., Jensen, D., Oates, T.: Efficient progressive sampling. In: Proceedings of the Fifth ACM SIGKDD International Conference on Knowledge Discovery and Data Mining, pp. 23–32 (1999)

[POSG04]   Park, B.-H., Ostrouchov, G., Samatova, N.F., Geist, A.: Reservoir-based random sampling with replacement from data stream. In: Proceedings of the SIAM International Conference on Data Mining (SDM'04), pp. 492–501 (2004)

[Sha02]    Shasha, Y.Z.D.: Statstream: Statistical monitoring of thousands of data streams in real time. In: Proceedings of 28th International Conference on Very Large Data Bases, pp. 358–369 (2002)

[Toi96]    Toivonen, H.: Sampling large databases for association rules. In: VLDB '96: Proceedings of the 22th International Conference on Very Large Data Bases, pp. 134–145 (1996)

[Vit85]    Vitter, J.S.: Random sampling with a reservoir. ACM Transactions on Mathematical Software 11, 37–57 (1985)

[YCLZ04]   Yu, J.X., Chong, Z., Lu, H., Zhou, A.: False positive or false negative: Mining frequent itemsets from high speed transactional data streams. In: Proceedings of the Thirtieth International Conference on Very Large Data Bases (2004)

[YSJ$^+$00]  Yi, B.-K., Sidiropoulos, N., Johnson, T., Jagadish, H.V., Faloutsos, C., Biliris, A.: Online mining for co-evolving time sequences. In: Proceedings of the 16th International Conference on Data Engineering, pp. 13–22 (2000)

[ZPLO96]   Zaki, M.J., Parthasarathy, S., Li, W., Ogihara, M.: Evaluation of sampling for data mining of association rules. In: Seventh International Workshop on Research Issues in Data Engineering, RIDE'97 (1996)

[ZWKS07]   Zhu, X., Wu, X., Khoshgoftaar, T., Shi, Y.: Empirical study of the noise impact on cost-sensitive learning. In: Proceedings of International Conference on Joint Conference on Artificial Intelligence, IJCAI (2007)

# Fast Loads and Queries

Goetz Graefe and Harumi Kuno

HP Labs, Palo Alto, CA 94304, USA
{goetz.graefe,harumi.kuno}@hp.com

**Abstract.** For efficient query processing, a relational table should be indexed in multiple ways; for efficient database loading, indexes should be omitted. This research introduces new techniques called zones filters, zone indexes, adaptive merging, and partition filters. The new data structures can be created as side effects of the load process, with all required analyses accomplished while a moderate amount of new data still remains in the buffer pool. Traditional sorting and indexing are not required. Nonetheless, query performance matches that of Netezza's zone maps where those apply, exceeds it for the many predicates for which zone maps are ineffective, and can be comparable to query processing with traditional indexing, as demonstrated in our simulations.

## 1 Introduction

In relational data warehousing, there is a tension between load bandwidth and query performance, between effort spent on maintenance of an optimized data organization and effort spent on large scans and large, memory-intensive join operations. For example, appending new records to a heap structure can achieve near-hardware bandwidth. If, however, the same data must be integrated into multiple indexes organized on orthogonal attributes, several random pages are read and written for record insertion. Some optimizations are possible, notably sorting new data and merging them into the old indexes, thus reducing random I/O but still moving entire pages to update individual records [GKK 01]. Nonetheless, load bandwidth will be a fraction of the hardware bandwidth.

As illustrated by Figure 1, the difference in query performance is just as clear. A fully indexed database permits efficient index-to-index navigation for many

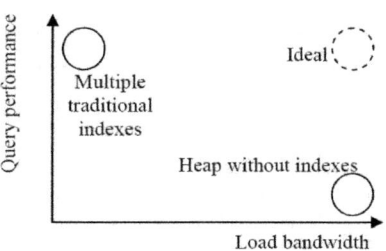

**Fig. 1.** The traditional trade-off between load bandwidth and query performance

A. Hameurlain et al. (Eds.): TLDKS II, LNCS 6380, pp. 31–72, 2010.

queries, with all memory available for the buffer pool. With index-only retrieval, non-clustered indexes may serve as vertical partitions similar to columnar storage. With a few optimizations, index searches can be very fast. For example, IBM's Red Brick product supports memory-mapped dimension tables. As less extreme examples, the upper levels of B-tree indexes can be pinned in the buffer pool and augmented with memory addresses of all child pages (or their buffer descriptors) also pinned in the buffer pool, and auxiliary structures may enable interpolation search instead of binary search. If, on the other hand, the load process leaves behind only a heap structure with no indexes, query processing must rely on large scans, sorts, large hash joins, etc. Shared scans may alleviate the pain to some degree, but dynamic sharing of more complex intermediate query results has remained a research idea without practical significance.

The tension between fast loads and fast queries is exacerbated by emerging applications, such as the short-term provisioning of databases in clouds and grids, where storage and computational resources might be occupied and vacated quite rapidly.

### 1.1    Prior Approaches

Two obvious approaches for addressing this problem immediately suggest themselves. First, one can improve the load bandwidth of indexed tables and databases, and in particular of B-tree indexes. Second, one can speed queries without indexes by improving scan performance, either by reducing the scan volume or by reducing the cost of scanning.

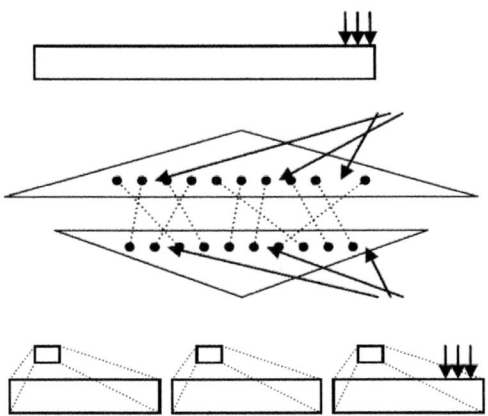

**Fig. 2.** Storage structures and load behaviors

Figure 2 illustrates these choices. A heap-organized table with no index is loaded by appending at the end, at the expense of large scans during query execution. A fully indexed table permits efficient query execution, at the expense of slow insertions at locations dictated by key values. A third option, originated by Moerkoette and later popularized by Netezza, Infobright, and others, is zone maps small summary structures created during loading that can guide query execution in many cases. These technologies serve as a foundation of the new techniques this article proposes and are described in detail below in Section 2.

## 1.2   Proposed Approach

This article explores how best to achieve efficient query processing in the presence of efficient database loading. First, it introduces new data structures called "zone filters" and "zone indexes" for zones or database segments, which go beyond prior work in enabling efficient query processing over heaps.

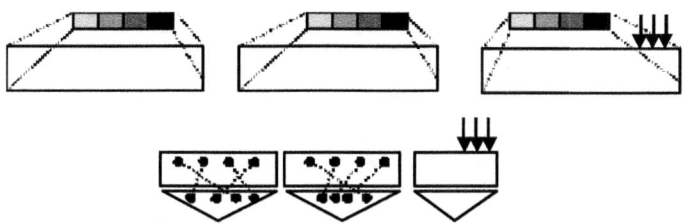

**Fig. 3.** Zone filters and zone indexes

As illustrated in the top half of Figure 3, zone filters are similar to zone maps, in that both summarize data that has been directly loaded into heaps. However, zone filters extend prior summarization data structures by providing richer and more general information about zone contents.

Zone indexes, shown in the lower half of the figure, embed in each zone or database segment informa-tion comparable to a traditional index. However, a zone index is limited in scope to the records within its zone or segment.

The second contribution of this article is to describe a new adaptive index optimization technique, called "adaptive merging," which exploits partitioned B-trees to enable the on-demand completion of partially formed indexes. The idea of adaptive merging is that data is loaded directly onto disk without sorting or indexing. Indexes are then created incrementally and automatically, as a side-effect of query processing.

Figure 4 shows initial index creation as a side-effect of query processing. When an un-indexed column is queried for the first time, the scanned data is sorted in memory and the resulting runs are saved as B-tree partitions. The size of these runs can be optimized to reflect the amount of available memory or other system resources, minimizing the impact of index initialization.

The initial index is optimized as a side-effect during subsequent queries. Figure 5 sketches how as a side-effect of processing subsequent queries, keys from these partitions are merged to create a new, unified, partition. This final merge par-tition is represents a full index of previously-queried keys. The more frequently a range is queried, the more rapidly that range will be fully merged. If only a

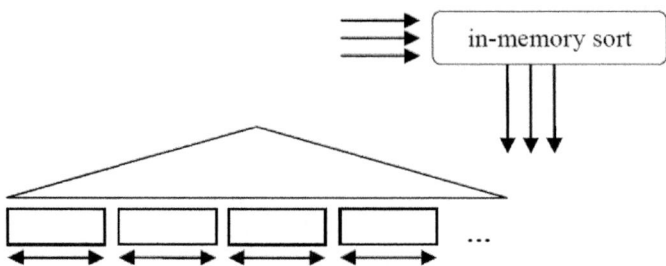

**Fig. 4.** Adaptive merging creates index partitions as a side effect of query processing

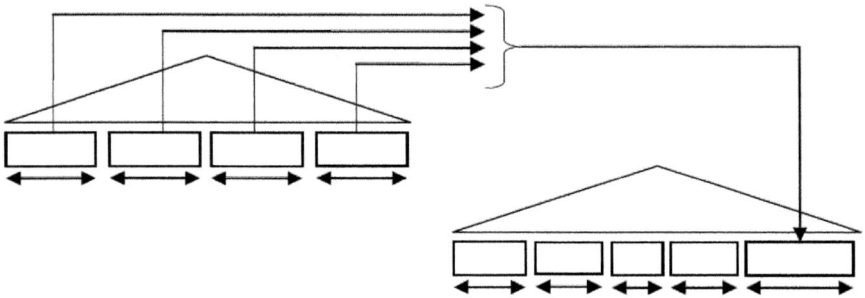

**Fig. 5.** Adaptive merging refines index partitions as a side effect of query processing

subset of the full range of keys is queried, then only that queried range will be merged. Queries that can leverage the final merge partition enjoy performance identical to queries against a full index. Intermediate queries enjoy performance benefits proportional to the degree of merging completed.

Finally, as a third contribution, this article introduces partition filtering, an approach that integrates the advantages of zone filters, zone indexes, and adaptive merging to provide a system that achieves optimal load bandwidth, optimal query processing performance, immediate availability of newly loaded database information, and yet requires only moderate code changes to B-tree indexes and their algorithms. In the proposed design, loading writes each item to disk only once and never revisits items with multi-pass algorithms such as external merge sort or B-tree reorganization. Nonetheless, query processing immediately after loading is as efficient as with zone filters and zone indexes. Compared to traditional B-tree indexes, the data structure left after loading and used by query processing is less refined. The advantage of the technique is that there is no need for index tuning, yet query processing can be almost as efficient as in a fully indexed database, as demonstrated in our experiments. Any reorganization after the load is complete is purely optional, and can be carried out incrementally and adaptively as a side effect of query processing.

We distinguish between mechanisms for achieving fast loads and fast queries and policies that can utilize such mechanisms. In this paper, we explore mechanisms their design, capabilities, and performance. We do not address questions of physical database design decisions such as index selection, although the techniques we discuss here complement such policies.

Section 2 begins by discussing the strengths and limitations of prior and related work. It divides these efforts into techniques that speed scans on data loaded without indexes and those that enable index construction to be carried out incrementally and online, after data has been loaded without indexes. Sections 3 through 5 introduce the designs for zone filters, zone indexes, adaptive merging, and partition filters, which allow us to load data without indexes, then incrementally and adaptively create and refine partitions, summaries, and indexes in response to the database workload. A performance evaluation in Section 6 is followed by Section 7, which offers a summary and conclusions from this effort.

## 2    Prior Work

This work draws on and generalizes several prior research efforts and results, reviewed and discussed below. It begins with prior attempts to build summary structures to guide searches over heaps of records. Next, it considers alternative methods for improving the physical layout of data so as to improve search times without the use of indices. Finally, it discusses prior efforts related to adaptive indexing.

### 2.1    Loading into Heaps

One approach to achieving fast loads and queries is to load data directly into physically separate unsorted heaps, then create small summaries of those heaps that can be used to guide, or even answer, queries. Fast queries are then achieved by avoiding scans of heaps that don't contain data relevant to a given query. These solutions differ in terms of the nature of the summary information.

**Small materialized aggregates.** Moerkotte [M 98] seems to have been the first to suggest parsimonious use of materialized views for query answering and for efficiency in query execution plans. The latter is based on a correlation of load sequence with time-related data attributes. Obvious examples include attributes of type "date" or "time"; less obvious examples with nonetheless strong correlation include sequentially assigned identifiers such as an order number, invoice number, etc. For example, if most or even all orders are processed within a few days of receipt, references to order numbers in shipments are highly correlated with the time when shipments occurred and were recorded in the database.

For those cases, Moerkotte proposed small materialized aggregates (SMAs) defined by fairly simple SQL queries. For example, for each day or month of shipments, the minimal and maximal order number might be recorded in a materialized view. Based on this materialized view, a query execution plan searching for a certain set of order numbers can quickly and safely skip over most days or months. In other words, the materialized view indicates the limits of actual order numbers within each day or month, and it can thus guarantee the absence of order numbers outside that range.

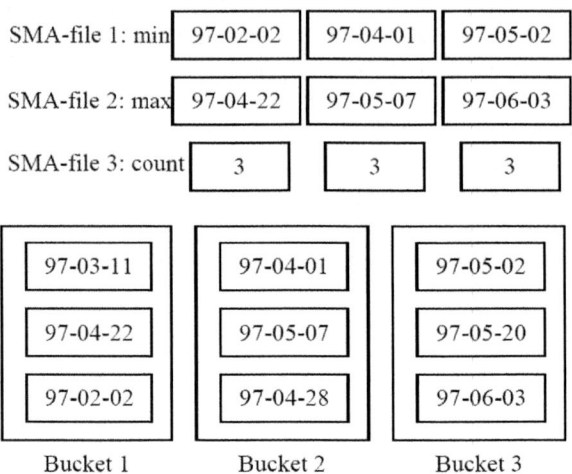

**Fig. 6.** Moerkotte's illustration

Figure 6 reproduces an illustration by Moerkotte [M 98]. It shows three buckets (equivalent to database segments or zones). Each bucket contains many records, with three records per bucket shown. Several small materialized aggregates are kept about each bucket, in this case three different aggregates, namely minimum and maximum values as well as record counts. The number of values for each of these aggregates is equal to the number of buckets. A query about a date range within the year 1998 can skip over all buckets shown based on the maximum date values in the appropriate file of small materialized aggregates.

Omission of maintenance during deletion results in possible inaccuracy, e.g., deletion of the last instance of the lowest or highest value. Such inaccurate aggregates cannot be used to produce accurate query results but can still be used to guide query execution plans. Specifically, they indicate the lowest or highest value that ever existed, not the lowest or highest value that currently exists. Thus, they can guarantee that values outside the range currently do not exist, even if the current values may not span the entire range.

In addition to minimum and maximum values, Moerkotte suggested supporting counts of records as well as sums of values. Both insertions and deletions can maintain those accurately and efficiently.

**Netezza.** Netezza employs a closely related technique that has proven itself in practice, both during proof-of-concept benchmarks and in production deployments. In order to simplify the software and to speed up query processing, only

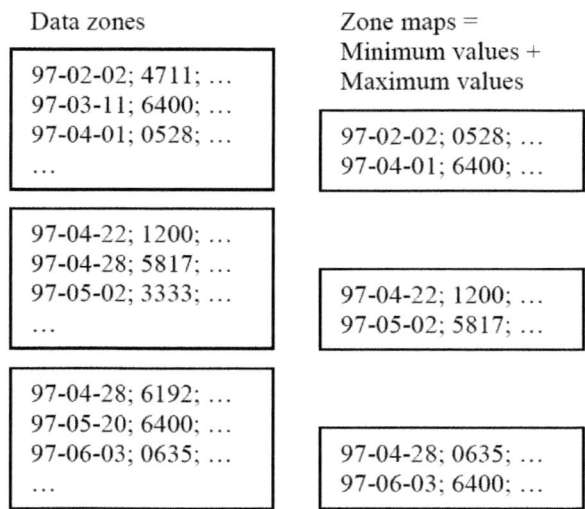

**Fig. 7.** Data zones and their zone maps

one record type is permitted within each database segment or "zone" (as far as we know). To avoid the need for manual tuning, zone maps cover all columns in a table. For each column and each physical database segment (e.g., 3 MB of disk space), minimal and maximal value are gathered during loading and retained in the database catalogs. During insertions into a database segment, the values are updated accurately. During deletion, maintenance of these values is omitted.

Query processing in data warehousing often involves query predicates on a time attribute. Often, there is a high correlation between the load sequence and data attributes capturing a time dimension, e.g., the date of sales transactions. In those cases, the information in the zone maps enables table scans to skip over much of the data. Reducing a scan by 99speed-up by factor 100.

Columns without correlation and predicates on those columns do not benefit from zone maps. For example, if there is a query predicate on product identifier, it is unlikely that the range of product identifiers within a database segment is substantially smaller than the entire domain of product identifiers. Thus, there will rarely be an opportunity to combine zone map information about multiple columns.

Figure 7 illustrates database segments or zones and their associated zone maps. Each zone contains many data records, all of the same type. Columns for data and for product identifiers are shown; additional columns and records are represented by " The associated zone maps contain precisely two records of the same type, one with minimum values and one with maximum values. (Figure 7 shows these as records of the same type as the data records, although a column-by-column representation is equally possible.) A query for the month 97-03 could readily skip over 2 of these 3 zones after merely inspecting the zone maps.

The date ranges of neighboring zone maps may overlap. As there usually is a strong correlation between load sequence and dates, these overlaps are limited in practice. The ranges of product identifiers overlap between zone maps; in fact, there does not seem to be any correlation between product identifiers and date or load sequence. Therefore, a range query on the date column likely benefits from the zone maps, whereas a range query on product identifiers typically does not. After zone maps reduce the number of zones that must be inspected in detail, Netezza relies on hardware support (FPGAs) to scan those zones and all their records quickly. While it is quite possible that automatic compilation from query predicates to appropriate FPGA programs is possible and remains robust through hardware generations, we believe that Boral and DeWitt's argument [BD 83] still holds that favors portable software running on the latest generation of general-purpose hardware. Thus, we propose a data organization within zones (or equivalent data segments) that enables efficient loading as well as efficient query processing.

With two extra records per zone, the space overhead of zone maps is negligible compared to the traditional design of multiple non-clustered indexes. While not a compression technique in the classic sense, this reduction in overhead saves cost and effort for purchase, power, backup, consistency checks, etc.

**Kognitio.** From very sparse information [M 08], it appears that Kognito's WX2 product supports compressed bitmaps per disk as its only indexing structure. They indicate which values appear in which disk blocks of 32 KB. Scans use them to skip blocks. In that sense, these bitmaps are similar in purpose to Netezza's zone maps. These bitmap indexes appear to support only equality predicates but not range predicates [M 08].

**Brighthouse.** The Brighthouse database management system by Infobright employs database segments called "data packs" [SWE 08]. Each data pack contains 64 K rows, stored column-by-column with compression but without sort order. For each data pack, a "data pack node" indicates the number of records and, for each column, minimum and maximum values, sum, and the count of Null values. In addition to data pack nodes, "knowledge nodes" may be created as side effect of query processing and contain equiwidth histograms with a single bit per range and 1,024 ranges between minimum and maximum values, character maps for string values, or join information that links pairs of data packs more like a page connectivity graph [MKY 81, MR 93] than a traditional join index [V 87].

The data pack nodes are quite similar to small materialized aggregates and to zone maps. The histograms in knowledge packs are similar to bit vector filters found in WX2. Range partitioning in the equiwidth histogram focuses on range predicates, although perhaps to the detriment of equality and "in" predicates. Moreover, range partitioning is liable to deteriorate in the presence of data skew or outliers. In contrast, the design proposed below provides mechanisms to eliminate the effects of outliers and skew, to support equality predicates by hashing values to bit positions, and to enable efficient search within database segments with minimal burden on load processing.

**Summary of loading into heaps.** By storing data directly into heaps partitioned by load order and then providing summaries of these heaps to guide queries, SMAs, Netezza, WX2, and Brighthouse can significantly reduce the amount of data scanned when answering queries without having to create or maintain indexes at load time. Avoiding the overhead of indexing maximizes load bandwidth, there is no need for complex index tuning, and scans for typical queries are very fast.

However, because all data remains stored in load order, partitioning into heaps only benefits query predicates on columns correlated with the load sequence. Furthermore, because search within a heap requires a scan, even if the queried column does correlate with the load sequence, a single outlier value that falls into the wrong heap can require that entire heap to be scanned.

## 2.2   In-Page Organization

Another approach to achieving fast loads and queries is to load data directly into heaps, possibly without indexing, and then leverage efficient physical layouts to speed query processing. Efficient in-page layout can improve query processing speed in two, orthogonal, ways: first, by reducing the amount of data scanned from disk and second, by optimizing use of cache lines once data has been read from disk. The techniques proposed in this article are compatible with both of these strategies.

Vertically partitioning data in the style of column stores may avoid retrieving unnecessary columns from disk, thereby reducing the amount of unnecessary data scanned when processing a query. Similarly, compressing B-trees (using techniques such as prefix or suffix truncation) potentially improves effective scan bandwidth to the point that even queries upon non-key attributes are fast without secondary indexes [BU 77]. Saving storage space by compression may increase both the number of records per leaf and also the fan-out of internal nodes. However, the approximately 10x improvement in scan bandwidth offered by column stores and 4x improvement gained by compression, even combined for a 40x improvement over a straight scan, do not begin to approach the performance of answering a point query using an index.

Other research efforts focus on optimizing use of cache lines, for example by caching keys frequently used in binary search [GL 01, L 01], organizing those keys as a B-tree of cache lines [RR 00], or dedicating "mini-pages" to the individual fields of the records stored on the page [ADH 01, G 08]. These techniques improve query performance significantly, notably for traditional row stores, by making the processing of in-memory data more efficient. This line of research is therefore particularly interesting in the context of zones, where an entire database segment or zone is read and processed as a single unit in memory.

## 2.3   Shared Scans

Yet another approach to improving scan performance is to coordinate query plans such that the underlying tables and indexes are scanned as few times as possible. This shared scan approach attempts to minimize the number of disk seeks [CDC 08, F 94, LBM 07, ZHN 07, ZLF 07]. In principle, shared scans and scans that skip over needless data are orthogonal techniques. However, inasmuch as the scan volume might be reduced due to skipping, less sharing may result. Thus, skipping needless data may reduce the benefits of shared scans. On the other hand, coordination among overlapping scans remains a valid and useful technique and should be applied as much as possible.

The remaining discussion in this article ignores the effects of concurrent scans and their coordination, instead continuing as if only a single scan matters. In an implementation destined for production use, techniques can be combined as appropriate.

## 2.4   Indexes Optimized for Loading

The prior approaches discussed thus far achieve fast loads by omitting indexes and then speed queries by reducing the amount of data scanned and/or by optimizing its processing. An alternative approach delays index creation and optimization until after loads have completed. The assumption is that loads are thus unencumbered by indexes, yet sometime later queries will benefit from the presence of indexes. The challenge is how to manage the overhead of building indexes while query processing. In this section, we discuss three approaches to delayed indexing: tree indexes with buffers, partitioned B-trees, and database cracking.

**Tree indexes with buffers.** Techniques optimized for efficient bulk insertions into B-trees can be divided into two groups. Both groups rely on some form of buffering to delay B-tree maintenance. The first group focuses on the structure of B-trees and buffers insertions in interior nodes [JDO 99]. Thus, B-tree nodes are very large, are limited to a small fan-out, or require additional storage "on the side." The second group exploits B-trees without modifications to their structure, either by employing multiple B-trees [MOP 00] or by creating partitions within a single B-tree by means of an artificial leading key field [G 03], as discussed next. In all cases, pages or partitions with active insertions are retained in the buffer pool. The relative performance of the various methods, in particular in sustained bandwidth, has not yet been investigated experimentally.

Figure 8 illustrates a B-tree node that buffers insertions, e.g., a root node or a parent of leaf nodes. There are two separator keys (11 and 47), three child pointers, and a set of buffered insertions with each child pointer. Each buffered insertion is a future leaf entry and includes a bookmark in Figure 8. The set of buffered insertions for the middle child is much smaller than the one for the left child, perhaps due to skew in the workload or a recent propagation of insertions to the middle child. The set of changes buffered for the right child in Figure 8. The figure includes not only insertions but also a deletion (key value 72). Buffering deletions is viable only in non-clustered indexes, after a prior update of a clustered index has ensured that the value to be deleted indeed must exist in the non-clustered index.

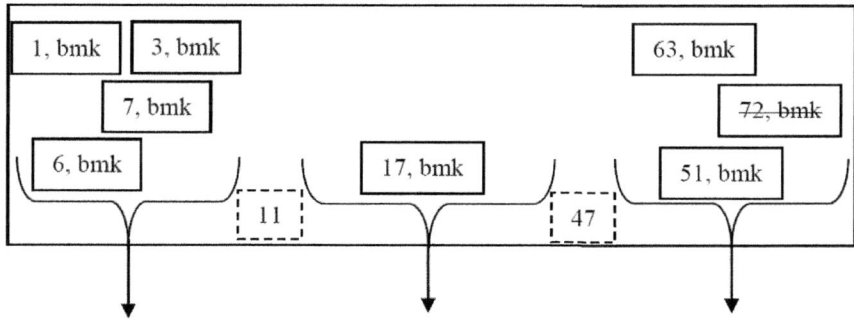

**Fig. 8.** Buffering in an interior B-tree node

**Partitioned B-trees.** Partitioned B-trees represent a specific tradeoff between database preparation effort during load processing and search effort during query execution [G 03]. One of the design goals was to combine loading at hardware bandwidth and query processing using a fully indexed database. Partitioned B-trees differ from traditional partitioning of tables and indexes by the fact that partitions are indicated using an artificial leading key field in the index records. In most index pages, this field is stored only once using well-known prefix truncation [BU 77]. Partitions are not reflected in the database catalogs. Thus, partitions are created and removed efficiently and without schema change simply by insertion or deletion of records with specific values in the artificial leading key field.

Loading adopts the logic for run generation from external merge sort and extends each B-tree being loaded with sorted partitions. Those are separated by an artificial leading key field containing partition identifiers. Incremental index reorganization may exploit idle times between loading and query processing or may be a preparatory side effect of query processing. The required logic is the same as merging in a traditional external merge sort.

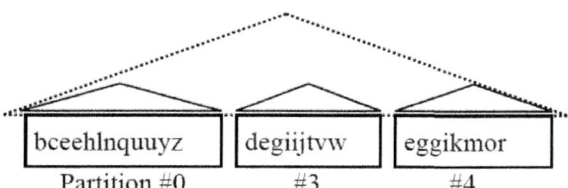

**Fig. 9.** Partitioned B-tree

Figure 9 illustrates a partitioned B-tree with 3 partitions, labeled 0, 3, and 4. B-tree reorganization into a single partition uses the merge logic of an external merge sort. Index search prior to complete index optimization needs to probe all existing partitions just like a query with no restriction on a leading B-tree column [LJB 95]. If a partition is extremely small, it might be more efficient to scan all

its records than to probe it using standard B-tree search. B-tree optimizations
for prefix and suffix truncation [BU 77] make it likely that partition boundaries
coincide with leaf boundaries.

If multiple indexes exist for a table being loaded, one can either sort each
memory contents once in each desired sort order and append the sorted records
as a new segment to the appropriate indexes, or one can append the new data
to only one index, say the clustered index of the table. In the latter case, the
load process can achieve near-hardware bandwidth. On the other hand, the incremental optimization steps not only move data from the appended segments to
the main segment but also propagate the insertions to the table's non-clustered
indexes that had been neglected during the initial load.

Partitioned B-tree indexes support not only roll-in (loading, bulk insertion,
data import) but also roll-out (purging, bulk deletion). First, an incremental task
moves the appropriate records to separate partitions; then, the actual deletion
step removes entire partitions and thus entire pages from the B-tree structure.
Other usage scenarios for partitioned B-trees include deferred index maintenance
as mentioned above, deferred maintenance of materialized views, deferred verifi-
cation of integrity constraints, deferred and incremental maintenance of statistics
such as histograms, separation of hot and cold index entries such that hot ones
can be cached in memory or in flash with minimal space requirements, and more.
These benefits could conceivably also apply to the database organization advocated below, forfeiting traditional clustered and non-clustered indexes in favor
of zone filters and zone indexes.

**Database cracking.** "Database cracking"[1] has pioneered focused, incremental,
automatic optimization of the representation of a data collection — the more
often a key range is queried, the more its representation is optimized for future
queries [KM 05, IKM 07a, 07b, 09]. This optimization is entirely automatic. In
fact, it is a side effect of queries over key ranges not yet fully optimized.

Column domain and storage array

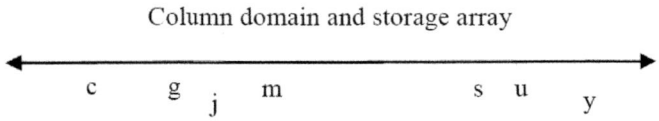

**Fig. 10.** Column store partitioned by database cracking

For example, after the column store illustrated in Figure 10 has been queried
with range boundary values $c$, $g$, $m$, $s$, and $u$, all key values below c have been
assigned to storage locations to the far left, followed by all key values between
c and g, etc. When a new query with range boundaries $j$ and $y$ is processed,
the values below g are ignored, the values between g and m are partitioned with
pivot value $j$, the values between m and u are returned as query results without

---

[1] Much of our description of the techniques in Sections 2.3 and 2.4 is from our earlier
papers [GK 10a-c].

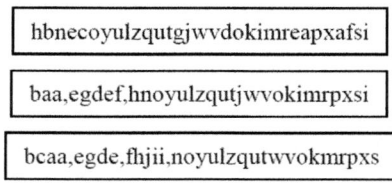

**Fig. 11.** Partitioning in a cracker index

partitioning or sorting, and the values above $u$ are partitioned with pivot value $y$. Subsequent queries continue to partition these key ranges.

Database cracking combines features of both automatic index selection and partial indexes. When a column is used in a predicate for the first time, a cracker index is created by copying all data values in the appropriate column from the table's primary data structure. When the column is used in the predicate of another query, the cracker index is refined as long as a finer granularity of key ranges is advantageous.

The keys in a cracker index are partitioned into disjoint key ranges and unsorted within each. As illustrated in Figure 10, each range query analyzes the cracker index, scans the key ranges that fall entirely within the query range, and uses the two end points of the query range to further partition the appropriate two key ranges. Thus, in most cases, each partitioning step creates two new sub-partitions using logic similar to the partitioning step in quicksort. A range is partitioned into 3 new sub-partitions if both end points fall into the same key range. This happens in the first partitioning step in a cracker index (because there is only one key range encompassing all key values) but unlikely thereafter [IKM 07a].

Figure 11 illustrates how database cracking refines the array of key values in a cracker index. Each character represents a record; the character is its key value. In this small example, each vowel appears twice and each consonant once. The top box shows the initial cracker index immediately after copying. The center box shows the cracker index after a range query for keys d through g. Both end points are included in the new center partition. The bottom box shows the partitions after a subsequent range query for key range f through j. The two new boundary keys are used to further refine two partitions. Partition sizes are very unevenly distributed.

Key ranges never queried are never partitioned or optimized. On the other hand, each individual data record is moved many times during the incremental transformation from the unoptimized initial representation to the fully optimized final representation. An evaluation of this weakness and a comparison with a data structure competing with cracker indexes is presented in [GK 10b].

**Summary of indexes optimized for loading.** Tree indexes with buffers, partitioned B-trees, and database cracking delay index creation and optimization until after loads have completed. The advantage is that loads are unencumbered by index construction or maintenance. The disadvantages are that queries are subject to the overhead of this delayed index construction, and that query performance does not benefit from indexes until some minimal number of queries has been processed.

# 3   Zone Filters and Zone Indexes

Beginning with this section, we first introduce proposals for zone filters and zone indexes, then describe adaptive merging in Section 4, and finally, in Section 5, introduce partition filters and describe how all of these innovations unite to form a powerful, integrated approach. The essence of this approach is to enable fast loads by loading data into initial partitions organized with minimal indexes and filters, and then enable increasingly fast queries by refining these indexes and filters incrementally, as a side-effect of query processing. The measure of success is both the overhead upon load processing compared to loading without any indexes and also the overhead upon query processing compared to querying with full indexes.

To this end, zone filters generalize small materialized aggregates, zone maps, and data pack nodes. Zone indexes permit indexed search within a zone. Adaptive merging enables the incremental build up of zone indexes, speeding search within a database segment. Partition filters bring the advantages of all of these techniques to the logical partitions of a B-tree.

## 3.1   Zone Filters

The design for zone filters differs from Moerkotte's small materialized aggregates by exploiting aggregates beyond those expressible in SQL. It differs from Netezza's zone maps in two ways: multiple low and high values and bit vector filters. It differs from Infobright's Brighthouse by employing multiple low and high values and by exploiting a bit vector filter with hashing rather than a rigid partitioning into 1,024 ranges of equal width. It differs from all three prior schemes by including both stored values and also simple arithmetic within rows.

**Multiple extreme values.** For each zone and each column, the design retains the m lowest and the n highest values. If $m = n = 1$, this aspect of zone filters equals zone maps. By generalizing beyond a single value at each end of the range, zone filters can be effective in cases for which zone maps fail, without substantial additional effort during loading.

If the Null value for a domain "sorts low," it may be the "lowest" value for a column in many zones. By retaining at least $m = 2$ low values, the lowest valid value is always included in the zone filter even in the presence of Null values. Thus, queries with range predicates can always test whether a zone indeed contains values in the desired range of values, and Null values do not create a problem or inefficiency.

In addition, retaining m+n extreme values per zone and column permits efficient query processing even in the presence of outliers. For example, if the column in question describes the business value of sales transactions, a single

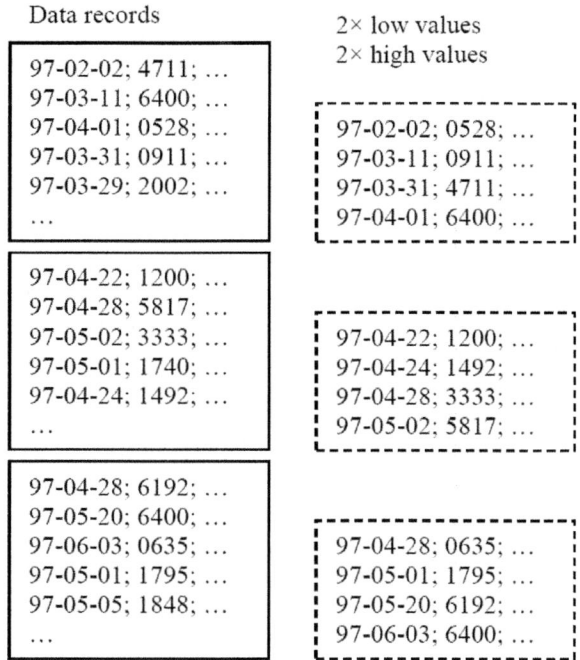

**Fig. 12.** Zone filter with m = 2 and n = 2

sales transaction of high value might greatly expand the range as described by the very lowest and the very highest value. Many sets of real-world measurements include outliers, not only values but also sizes, weights, distances, intervals, etc. Sometimes data analysis focuses on such extreme values, sometimes it focuses on the main population, e.g., the "bread and butter" customers. Even a few low and high values, ideally one more value than could be considered outliers, can ensure that query predicates can be handled effectively, i.e., only those zones are inspected in detail that contribute to the query result with high probability.

Figure 12 shows a zone filter (dashed outline) with m = n = 2 or four synopsis records per database segment or zone (solid outline). Minimum and maximum values are equal to those shown in Figure 7. Notice there is only a single data record for month 97-02, which may therefore be considered an outlier. For a query with predicate "97-02-22" or "between 97-02-16 and 97-02-28," the zone maps in Figure 7 does not exclude that zone, whereas the zone filters in Figure 12 do based on their multiple low and high values.

If there are fewer than m+n distinct values, some of the m+n values in the zone filter might be Null. The lowest value actually found in the domain is always retained as the lowest value in the zone filter. In other words, the lowest value in the zone filter is Null if and only if the Null value occurs in the zone.

In such cases, or if the search key is smaller than the largest among the m lowest values or larger than the smallest among the n highest values, the set of extreme values in the zone filter supports not only range ("<", "between") predicates but also equality ("=", "in") predicates, even for query constants that are within the range between low and high values. For example, if the m = 3 lowest values retained in a zone filter are (4, 7, 12), a search for value 9 can safely skip the zone. Thus, in the case of domains with few distinct values, the generalization from Moerkotte's SQL aggregates and Netezza's zone maps to m+n extreme values not only handles Null values and outliers but also offers new functionality and performance improvements for an additional set of queries.

Some readers might fear the cost of maintaining m+n values, in particular for non-trivial values of m and n. In that case, the load process should employ two priority queues. The values are the roots of these priority queues are the mth lowest and the nth highest value seen so far. The priority queues are initialized with the first m+n distinct values in the load stream. Each subsequent value in the load stream is compared with these root values, and if necessary a pass through one priority queue is required with log2(m) or log2(n) comparisons. The two required comparisons are comparable to the 2 comparisons required while building a zone map.

**Bit vector filters.** If the values in a database column have no correlation with the load sequence of the table, the range between minimal and maximal actual value in each zone will approach the entire domain of the column. In other words, zone maps and even the generalization with m+n extreme values will provide hardly any reduction in the number of zones that need to be scanned in detail.

Each zone might contain only a few distinct actual values, or at least substantially less than the entire domain. For those cases, we propose to include bit vector filtering in the zone filter. For each zone and for each column, a bit vector filter provides a synopsis of the actual values, and scans with equality predicates can exclude zones where the constant literal in the query predicate maps to an "off" bit in the bit vector filter. The number of bits should substantially exceed the number of distinct values [B 70].

For example, consider a table for sales of seasonal products, or any other case where items are introduced and discontinued over time. Depending on how new product identifiers are assigned, zone maps may be effective for their highest values. Old products are usually discontinued in a more random numeric sequence, so zone maps would not help with the lowest values. Bit vector filters, on the other hand, are independent of the sort order of product identifiers, their introduction, and their discontinuance, whereas zone maps depend on correlation between column values and the load sequence.

Figure 13 illustrates a zone filter with a bitmap per column; the m+n records with low and high values are omitted. In this figure, the last decimal digit of the date or the product identifier is used as bit position. E.g., value 0528 in the second column maps to bit position 8 in the second bitmap within the zone filter.

As in bit vector filtering for other usage scenarios, a hash function should be used to map values to bit positions. Unless hash collisions are exceedingly frequent, the bit vector filters can be effective even if the bitmap size is moderate, e.g., a few hundred bytes.

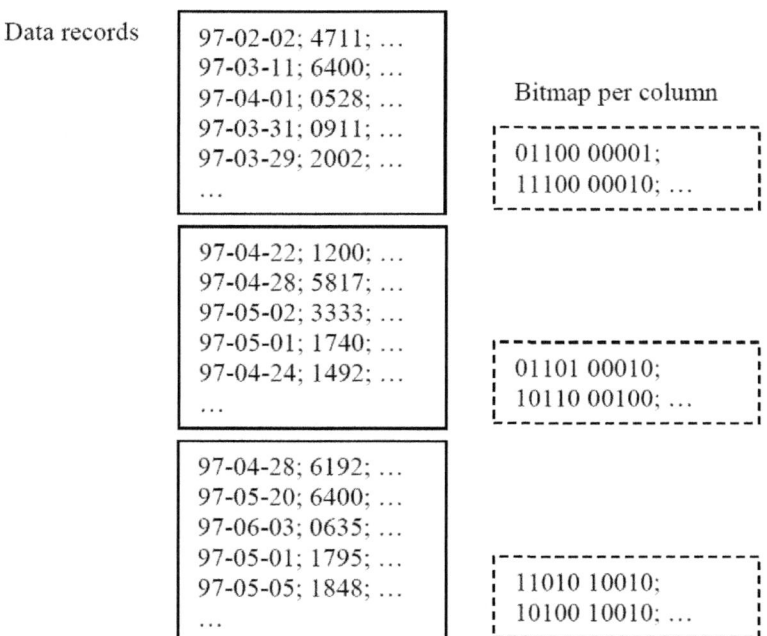

Data records

97-02-02; 4711; ...
97-03-11; 6400; ...
97-04-01; 0528; ...
97-03-31; 0911; ...
97-03-29; 2002; ...
...

Bitmap per column

01100 00001;
11100 00010; ...

97-04-22; 1200; ...
97-04-28; 5817; ...
97-05-02; 3333; ...
97-05-01; 1740; ...
97-04-24; 1492; ...
...

01101 00010;
10110 00100; ...

97-04-28; 6192; ...
97-05-20; 6400; ...
97-06-03; 0635; ...
97-05-01; 1795; ...
97-05-05; 1848; ...
...

11010 10010;
10100 10010; ...

**Fig. 13.** Zone filters with bitmaps per column

It might be useful to reserve one bit position in the bitmap for Null values, without the possibility of hash collisions. By doing so, queries specifically searching for missing values (Null values) can be processed very efficiently, with little impact on all other queries. If the bit vector filter indicates the presence of a Null value, there is no need to include it among the lowest m values in the zone filter. Thus, an additional actual value can be included in the zone filter for a slight increase in effectiveness.

As there is no need to capture occurrence of a particular value twice in a zone filter, and as the zone filter is the combination of extreme values and bit vector filter, the m+n extreme values are not represented in the bit vector filter. Thus, if no more than m+n distinct values occur in a zone (in one column), the bit vector filter is entirely clear.

Bit vector filters typically do not support range predicates very well. If a query range is small, it may be possible to enumerate the values in the range. E.g., the predicate "between 4 and 6" can be rewritten to "in (4, 5, 6)" for an integer domain. In order to solve the problem more generally, some bit positions in the bit vector filter may be dedicated to ranges. E.g., the column value 154 might be mapped to a bit position for the specific value as well as another bit position dedicated to the range 100-199. A query with the predicate "between 167 and 233" might be able to skip over many zones even if some data values such as 154 might create false positives to be eliminated by the detail scan.

These bit vector filters are really a special form of aggregate. Thus, as records are added to a zone during load processing, each record adds to the bit vector filter. They are different than traditional aggregates in that record deletions are not reflected in a bit vector filter. In that sense, a bit vector filter does not reflect the current actual values in a zone but the set of values that have existed in the zone and column since the bit vector filter was created. If deletions are frequent, it might be useful to recomputed bit vector filters every now and then in the affected database segments or zones.

**Derived values.** In addition to bit vector filters as well as m+n rows with low and high values (stored in row or column format), the proposed design includes similar information about derived columns. The goal is to guide scans with predicates against arithmetic expressions.

For example, if each row contains two dates (or other forms of points in time), the time interval between those is a likely candidate for query predicates. After all, time and timely performance are essential in business operations and therefore in business intelligence.

For location information, polar coordinates can be derived if rows contain Cartesian coordinates, and vice versa. Multiple locations lend themselves to distance calculations. Other examples for derived values include ratios and differences among numeric values, in particular those representing money and physical measurements such as size or weight.

While intervals between points in time are most obvious candidates for zone filters, other expressions may be observed in the workload. Note that it is not imperative that all database segments or zones have precisely the same zone filter information. As the information about the workload accumulates, more information may be captured for new database segments or zones than for existing ones. This is analogous to autonomic database systems summarizing tables using histograms as required by the actual query workload. Our proposal introduces the flexibility for individual zones rather than entire tables.

**Alternative aggregates.** For further generality, zone filters may include additional aggregations beyond low and high values and a bit vector filter. For example, Brighthouse "data pack nodes" contain sums and counts of Null values, which can easily become part of zone filters. Various moments (sums of powers) can be calculated and can provide instant information about variance, covariance, regression, correlation, etc.

Some designs for cardinality estimation during query optimization employ most frequent values captured in histograms; these, too, could be retained in zone filters although they would be local and thus have limited value in many cases. Moreover, maintenance of most frequent values during record deletion is inefficient and might actually be omitted such that the zone filter may become out-of-date with respect to frequent values. For both reasons, frequent values just like correlation and regression would more likely be useful for setting a priority among zones rather than filtering them outright.

Minimum bounding rectangles can also provide interesting information, in particular for spatio-temporal data and if the data within a zone can be captured in multiple small minimum bounding rectangles rather than a single minimum bounding rectangle (implied by the appropriate low and high values).

**Summary of zone filters.** Zone filters integrate techniques previously introduced in isolation as small materialized aggregates, zone maps, data pack nodes, etc. Moreover, they generalize them to handle outliers (using m+n low and high values instead of only the minimum and maximum values), queries on expressions (such as intervals between date and time values), and further statistics (e.g., minimum bounding rectangles and regression).

## 3.2   Zone Indexes

While zone maps, zone filters, and their like let scans skip over many parts of a table or database, zone indexes are a technique that enables efficient search within such a part. For example, if a zone filter indicates that a certain database segment may contain records that satisfy a query predicate, the zone index for that segment enables very efficient search within the zone without scanning the entire zone and all the records it contains.

The essence of zone indexes is to embed in each zone or database segment information comparable to a traditional index, limited in scope to the records within that zone or segment. Embedding zone indexes within the zone or database segment aids both load and query performance. While the segment is assembled in the buffer pool during loading, a zone index can be created within memory and without accesses to external storage. During query processing, any zone or segment not eliminated by the zone filter can be searched efficiently using this index, which is loaded into the buffer pool in memory together with the detail data in the segment. (It is also possible to load the zone index only at first and use it as an additional filter.)

The proposed default design for zone indexes is to index every column just like zone maps and zone filters. In that case, the data records may be sorted according to one column only. This is analogous to clustered and non-clustered indexes in traditional databases. The main difference is that every value needs to be stored only once, that all these indexes may share data values and that pointers (offsets) may be used instead of data values very much in the spirit of T-trees optimized for in-memory indexing [LC 86]. Moreover, the indexes may be optimized for CPU caches [GL 01, L 01]. The reason these techniques are applicable is that entire database segments are moved between memory and disk as a unit during database loading and during query processing.

Figure 14 illustrates three database segments, or zones. The left-most and right-most zones have records on the right and a clustered index on the left. The middle zone has records in the center, a clustered index on the left, and a non-clustered index on the right. Only one non-clustered index is shown although an index for each column or even column combination is possible. Sharing values between the records and the index nodes is omitted as it would contribute more

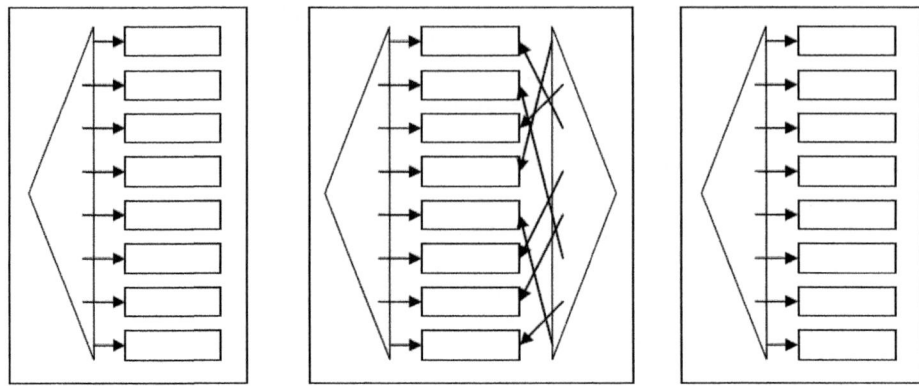

**Fig. 14.** Database segment with clustered and non-clustered indexes

confusion than clarity to the diagram. A data organization by column is similarly possible and might serve to reduce faults in the CPU cache during searches and it might enable more effective data compression.

There are many alternative forms of organizing and indexing records in a database segment or in a large page. In addition to traditional indexes, there are multi-dimensional indexes and multi-dimensional clustering. Moreover, just as the set of materialized aggregated may differ among database segments or zone, so may the physical organization. For example, the choice of clustered index may differ from one database segment or zone to another, and two columns may be indexed within multiple database segments or zones using a single two-column index or two single-column indexes.

Similarly, encoding and compression may differ from one database segment or zone to another, for example, due to a different set of distinct values in a column. The important points here are that indexes are applied within database segments used as contiguous disk storage and as in-memory data structures, they are created during high-bandwidth database loading with moderate processing and memory requirements, and storage formats may differ among database segments or zones.

Prior work has analyzed and improved the search performance within individual database pages. Such improvements promise to be even more benefit for entire database segments or partitions of several megabytes in size. For example, would a column organization within each segment improve query performance? More importantly, what can improve search performance beyond the simple linear search? Netezza relies on specialized hardware; a traditional argument suggests that software running on general hardware can out-perform special hardware, e.g., a B-tree index may be faster than a hardware-supported scan.

Like zone filters and unlike traditional indexes, zone indexes can be constructed quite inexpensively as part of the load process, with little processing effort and memory. Creation of a traditional index requires sorting the future index records; for a large index, an external merge sort requires external storage for intermediate runs and multiple passes over the data. Maintenance of such an index during a large load requires either many random insertions or sorting the change set followed by merging the change into the existing index. Partitioned B-trees do not reduce this effort, they merely create the ability to perform the work later and incrementally.

# 4    Adaptive Merging

Adaptive merging aims to combine efficient merge sort with adaptive and incremental index optimization so that we can load data without indexes, then create and refine indexes incrementally as a side effect of the queries in a workload [GK 10a, 10b, 10c]. Like database cracking, it requires a flexible underlying storage structure for partially and locally optimized index states. Partitioned B-trees appear to be an ideal choice.

The essence of partitioned B-trees, as described above in the section on prior work, is to use standard B-trees to persist intermediate states during an external merge sort, to provide efficient search at all times even before B-tree optimization is complete, and thus to separate run generation and merging into independent activities with arbitrary intermediate delay. Partitioned B-trees can also capture intermediate states during index creation, data loading, view materialization, etc.

The essence of adaptive merging is to exploit partitioned B-trees in a novel way, namely to focus merge steps on those key ranges that are relevant to actual queries, to leave records in all other key ranges in their initial places, and to integrate the merge logic as side effect into query execution. Thus, adaptive merging is like database cracking as it is similarly adaptive and incremental but they differ fundamentally as one relies on merging whereas the other relies on partitioning, resulting in substantial differences in the speed of adaption to new query patterns.

The differences in query performance are due to data being kept sorted at all times in a B-tree. The difference in reorganization performance, i.e., the number of queries required before a key range is fully optimized, is primarily due to merging with a high fan-in as opposed to partitioning with a low fan-out of 2 or 3. The following sections explain in more detail.

## 4.1    Index Selection

For index selection, the design copies the heuristic from database cracking: When a column is used in a predicate for the first time, a new index is created by copying appropriate values. Refinements such as external guidance which indexes to avoid and which ones to choose with priority in queries with multiple predicates, partial indexes, multi-column indexes, consideration of other predicates and their desirable indexes, etc. apply quite similarly to both techniques. The record formats are also similar unless compression is used, e.g., for duplicate key values.

The ordering of data records in an initial copy, however, is quite different due to partitioning in database cracking versus merging in the approach.

## 4.2   Initial Index Creation

The initial format of a partitioned B-tree consists of many partitions. Each partition is sorted, but the partitions most likely overlap in their key ranges. Subsequent merging brings the B-tree closer to a single sort sequence in a single partition, as described later.

The initial creation of a new partitioned B-tree performs run generation using an in-memory algorithm such as quicksort or replacement selection. The advantage of the latter is the opportunity for runs larger than the memory allocation during initial index creation. Each run forms a partition in the new B-tree.

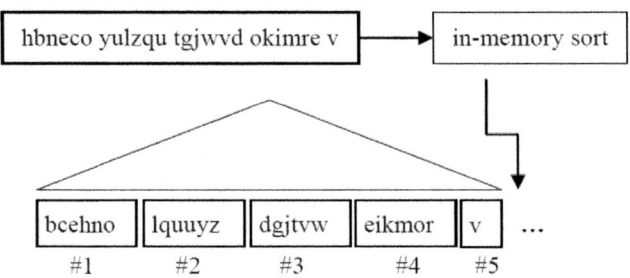

**Fig. 15.** Appending partitions during initial index creation

Figure 15 illustrates data movement during initial index creation. The upper box shows the input, entirely unsorted. The lower box shows the initial index, i.e., records and partitions within a partitioned B-tree. Run generation during copying produces runs of 6 records in this small example. When an un-indexed column is queried for the first time, a new partitioned B-tree is instantiated to hold its index. As the keys of the column are scanned from disk as part of routine query processing, each run's worth of data is sorted in memory and then appended to the B-tree as a new partition. Because they are stored in a B-tree structure, these partitions can now each be queried efficiently by subsequent queries. A run generation algorithm such as quicksort is used to append as many partitions as necessary. Their number depends primarily on input size and memory allocation but also on sort algorithm and any incidental correlation between the sort order in the data source and in the new index.

Search performance immediately after index creation depends on the count (and thus the average size) of the partitions in the partitioned B-tree, as does the break-even point between probing each partition with a traditional B-tree search and an end-to-end scan of the index. For example, if scan bandwidth is 100 MB/s and each probe takes 20 ms, partitions larger than 100 MB/s 20 ms = 2 MB ought to be probed rather than scanned, corresponding to a modest memory allocation of 1 MB during run generation by replacement selection. Note that the "lock footprint" can be smaller during probing than during scanning, further favoring probing over scans. Modern flash storage also favors probing over scans. Nonetheless, scanning is always possible if desired, e.g., in order to exploit shared scans.

**Incremental index optimization.** When a column is used in a predicate for the second time, an appropriate index exists, albeit not yet fully optimized and merged into a single partition. In this situation, a query must find its required records within each partition, typically by probing within B-tree for the low end of the query range and then scanning to the high end. Instead of just scanning the desired key range one partition at a time, however, the query might as well scan multiple partitions in an interleaved way, merge these multiple sorted streams into a single sorted stream, write those records into a new partition within the partitioned B-tree, and also return those records as the query result. The data volume touched and moved is that of the query result.

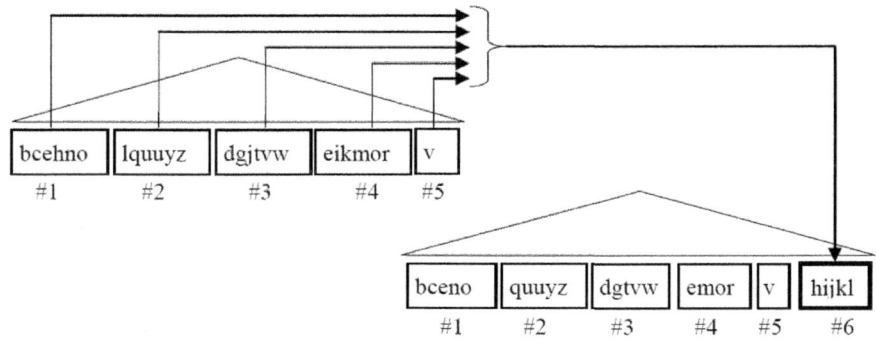

**Fig. 16.** Partitioned B-tree before and after a second query

Figure 16 continues the concrete example of Figure 15. It illustrates merging and data movement during a second query, a range query on keys $h$ through $l$. The top half shows the partitions that were generated as a side effect of the first query. In processing the second query, records satisfying the query predicate are automatically merged into the new partition, as shown in the bottom half of Figure 16. The final merge partition is marked by a thicker border. A subsequent query within key range $h$ through $l$ would access only partition #6.

Ideally, a single merge step suffices to merge records from all existing partition into a single, final partition. That is, if the number of initial partitions is smaller than the fan-in as limited by the memory allocation available for merging, then the query may leave the keys within its query range in a single location comparable to a traditional, fully optimized B-tree index. However, if more than a single merge step is required to transform the B-tree index from many initial partitions into a single final partition, each key range must be searched and merged by multiple queries before it is in its final, completely optimized format.

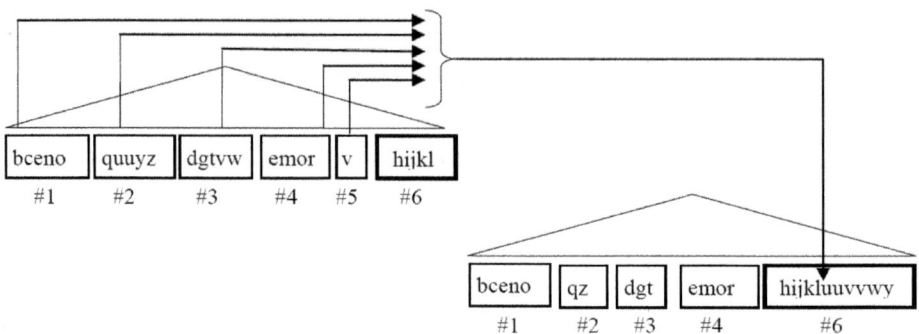

**Fig. 17.** Partitions merge as a side-effect of a query on key range $u$ through $y$

All subsequent queries must analyze their range predicates for overlap with prior queries and the merge effort they applied to the index. Queries whose key ranges are not contained within the merge partition(s) must probe all existing partitions, extract the required records, merge them, and move them to a new partition, as shown in Figure 17, below.

For example, Figure 17 shows the process of a fourth query, for key range $u$ through $y$. The top half shows the partitions left after processing the first three queries. In processing the fourth query, records satisfying the query predicate are automatically merged into the new partition, as shown in the bottom half of the figure. Note that as all remaining records were moved out of partition 5 into partition 6, and partition 5 was eliminated as a side effect of query processing.

Key ranges without query activity are never reorganized or merged. Those keys remain in the initial runs produced by run generation. Thus, as in database cracking, no effort is wasted on inactive key ranges after the initial copy step. By adaptively merging only those key ranges actually queried, and by performing merge steps as side effects of query execution, adaptive merging preserves the main strength of database cracking. The main difference is in the speed of adaptation, i.e., the number of times each record is moved before it is in its final location.

The number of merge steps for each key range is equivalent to the merge depth in an external merge sort, i.e., logF (W) for W initial runs merged with fan-in F. With the memory sizes of modern computers, sort operations with a single merge step are common, and sort operations with more than two merge levels are quite unusual. Just as in external merge sort with optimized merge patterns, the merge depth may not be uniform for all records and thus the average merge depth might be a fraction, e.g., 1.

In fact, the number of merge steps per record is a key difference between database cracking and adaptive merging. The merge fan-in can easily exceed 100, whereas the partitioning fan-out in database cracking is usually 2 or 3, limited by the number of new partitioning keys available in any one range query. Thus, database cracking may move each record many times before its final place is found. The exact number depends on the size of partitions to which no further cracking is applied and the size of the initial partitions in the proposed design.

## 4.3   Table of Contents

As in database cracking, an auxiliary data structure is required to retain information about reorganization efforts already completed. In fact, the set of keys is the same in the auxiliary data structures for database cracking and for adaptive merging. The information associated with those keys differs. In database cracking, the start position of the partition with the given is indicated. In adaptive merging, the data structure indicates the range of identifiers for partitions with records in the given key range.

For example, suppose that run generation creates runs with identifiers 1 through 1,000. All key ranges can be found in this set of partitions. After a key range has been merged once, say with merge fan-in 32, records within this key range can be found in partitions 1,001 through 1,032 but not longer in partitions 1 through 1,000. A key range merged twice can be found only in partition 1,033. Query performance in such key ranges equals traditional B-trees.

## 4.4   Concurrency Control, Logging, Recovery

As the proposed structure is a B-tree, even if an artificial leading key field is added, all traditional methods for concurrency control, logging, and recovery apply.

In addition, key prefixes could be locked, a generalization of Tandem's "generic locks" [G 07a]. When a conflict arises, a merge step can be committed immediately because merge operations do not change the contents of the index, only its representation.

The logging volume during merge operations can be reduced to allocation-only logging. In this mode of operation, the page contents are not logged during merge steps, neither deletions in the merge inputs nor insertions in the merge output. Deletion of individual records can be implemented as updates from valid records to "ghost" records (also known as pseudo-deleted records). A single small log record suffices for multiple records. Deletion of entire pages can be captured by a single small log record. Insertion of new pages requires that the new pages be flushed to disk before the data sources for the page contents are erased, i.e., before committing a merge step.

## 4.5   Updates

In this section, we briefly outline alternative techniques for insertions, deletions, and record modifications. In each case, the first technique is similar to traditional techniques whereas the second one is optimized for efficient completion of many small transactions.

Insertions can be either placed into the final target partition or gathered in a new partition dedicated to gathering insertions. This partition ought to remain in the buffer pool such that all insertions only update in-memory data structures (other than the recovery log). Multiple new partitions may be added over time.

Deletions can either search for the appropriate record in the index, in whatever partition it might be found, or they insert "anti-matter" quite similar to the "negative" records employed during maintenance of materialized views and during online index creation.

Modifications of existing records can be processed either as traditional updates after an appropriate search or they can be processed as pairs of deletion and insertion, with alternative processing techniques as outlined above.

If insertions, deletions, or updates create new partitions in a B-tree, i.e., introduce new partition identifiers, those partitions and their key ranges must be reflected in the table of contents such that subsequent queries search in all appropriate partitions.

## 4.6   Variations

Several variations and optimizations are possible beyond the design described so far. This section lists some ideas; we have not yet analyzed them for their true practical value or precise performance effects.

First, the basic idea seems well suited to capturing and indexing continuous streams, in particular if multiple independent indexes are desired for a single stream. Incoming records are always appended to all indexes in partitions formed by run generation. Continuous "trickle updates" in data warehouses are a special case of streams that can be indexed using the proposed techniques. Partitioned B-trees with adaptive merging also promise advantages for concurrency control, logging, and recovery in such environments.

Second, the general technique applies not only to disk-based databases but also to databases on flash devices and even to in-memory databases. The resulting differences are quantitative rather than qualitative. For example, due to very fast access latency, smaller page sizes are optimal for flash devices, resulting in higher merge fan-in with a fixed memory allocation and thus fewer merge levels from initial runs to a final, fully optimized B-tree [G 07b]. For in-memory databases, optimization of cache faults leads to run generation within the cache and explicit merge steps to form memory-sized indexes [NBC 95]. These concepts can be combined using self-similar algorithms and data structure to form cache-size in-memory runs, merge those to form initial memory-sized runs on flash, and finally merge those to form initial runs on disk.

Third, partitioned B-trees are useful not only for efficient search but also for efficient query execution with merge joins, "order by" clauses, etc. The final merge activity in the query is precisely equivalent to B-tree optimization, and the merge output can replace the previous partitions with a single, fully optimized partition. For orderings on B-tree fields other than the leading field, a general mechanism comparable to MDAM [LJB 95] seems possible but has not yet been described in the literature.

Fourth, adaptive merging in combination with partitioned B-trees provides mechanisms for dynamically adjusting query costs for the purpose of workload management. During index creation, it is possible at any time to defer the remaining key range within the data source. Doing so speeds up the current query but leaves the new index only partially populated. During index optimization, it is possible at any time to reduce the fan-in of merge steps or to interrupt all merge activity in order to defer some merge effort to later queries. Doing so frees up memory (merge input buffers) and speeds up the current query but fails to optimize the key range for subsequent queries.

Fifth, B-tree optimization and partition merging does not depend on queries. Instead, any idle capacity can be used to optimize a partitioned B-tree for future queries. Adaptive merging can focus on those key ranges that have been queried once but are not yet fully optimized. Database cracking, in contrast, cannot exploit prior queries during idle times because it requires a new partitioning key for each additional step.

Finally, instead of merging the precise key range of a query, the logic could be modified to consume entire B-tree leaves. Space management would become simpler and more efficient, whereas the table of contents would become more complex. Consequently, determining the required partitions during query execution would also be more complex. As a compromise, one can extend a query range to the next "short enough" separator key, quite similar to the key optimizations in suffix truncation (compression) [BU 77]. For example, if the query range starts with "Smith," the merge could start with "Sm". Even an equality query could merge an appropriate key range, for example all keys starting with "Sm". If suffix truncation is applied during B-tree construction, the probability is high that merge range coincides with boundaries between leaf pages in all input partitions. In fact, such a policy might be very useful to avoid an excessive number of small merge steps and thus to ensure efficient adaptation of an index to a new query pattern. If multiple merge levels are required, the heuristics might differ among the levels in order to avoid repeatedly searching a large number of initial partitions. The experiment below extends each merge range in both directions to a multiple of the largest power of two smaller than the width of the query range.

## 5   Partition Filters

Just as zone maps and zone filters summarize the contents of unordered heaps, partition filters summarize the contents of partitions. They employ the same techniques such as m+n low and high values and bit vector filtering. They thus enable up-front identification of partitions that do not contain any records contributing to a given query's result. Skipping over these partitions can dramatically reduce the merge effort of adaptive merging. By minimizing the overhead imposed on query processing, partitioned B-tree indexes with adaptive merging and partition filters promise performance equal to zone filters prior to index optimization and performance equal to a fully index database after index optimization.

Partitions based on data values might be used in many ways, and partition filters can be combined with all of them. Partition filters can be combined with range partitioning, hash-partitioning, or combined schemes such as hash-partitioning small ranges in order to minimize latency for small range queries and maximize bandwidth for large range queries. In spite of this generality, our immediate interest is partition filters applied to the partitions in a partitioned B-tree.

This section explores how partitioned B-trees, partition filters, and adaptive merging complement each other. Section 5.1 explains how partitions and partition filters enable fast loads. Section 5.2 discusses how partition filters plus adaptive merging achieve fast query processing. Finally, Section 5.3 explores some potential future improvements to an adaptive partition filter system.

## 5.1   Loading

The essence of data loading in the partition filter approach is deliberately simple: as data flows from source to disk, two side effects occur. First, as with adaptive merging in a partitioned B-tree, new data is sorted into runs using a workspace in memory; these initial runs are appended to the B-tree as new partitions. Second, as with zone filters in a heap, a partition filter is created for each new partition.

The key to fast load performance is that partitioning logic can be adjusted to fit resource availability, system load, or data characteristics. The most obvious adjustment is to ensure that each new partition can be sorted within the available memory. A smaller workspace reduces not only memory needs but also CPU effort per record due to fewer comparisons during run generation.

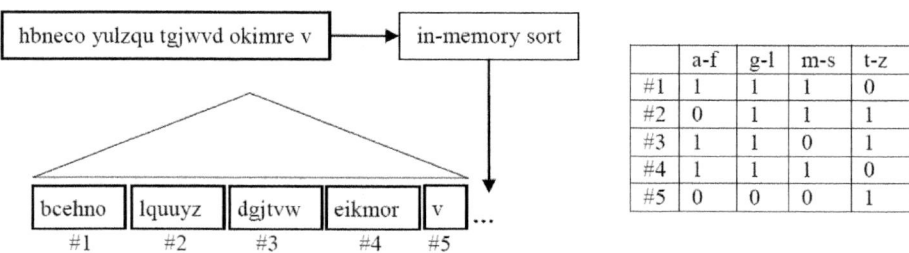

**Fig. 18.** Data is loaded into partitions, and bitmap from partition filters

Figure 18 extends the previous running example from Figure 17. Although partition filters include other information about partition contents, this figure shows only one component the bit vector filter for the indexed column indicating the ranges of values each partition contains. For example, each row in the table on the right is a partition filter for one partition and one column in the index; the last row indicates that all values in that partition are in the range $t - z$. Like a zone filter, a partition filter's contents can easily be calculated while performing the initial in-memory sort.

This combination of partitioned B-trees, adaptive merging, and partition filters is deliberately kept as simple as possible. As discussed in [G 03], partitioned B-trees can be implemented entirely using existing infrastructure of B-tree structures, with require minimal code changes to traditional B-tree indexes and their algorithms. Yet, as described below, the combination of partition filters and the adaptive merging of B-tree partitions enables the efficient incremental creation and refinement of indexes to reflect actual workload.

## 5.2   Query Processing

Partition filters complement adaptive merging by indicating which partitions need to be merged and which ones can be skipped. Partitions filters with low and high values and with range-based bit vector filtering benefit queries with range predicates, whereas partition filters with hash-based bit vector filtering benefit queries with equality predicates. Thus, in order to benefit all queries, partition filters must include multiple types of summary information for each partition.

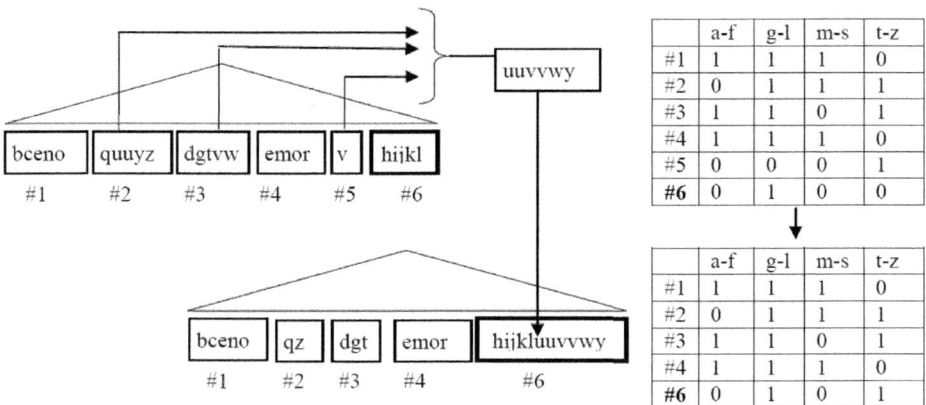

**Fig. 19.** Partition filter indicates partitions that do not need to be merged

For example, Figure 19 extends Figure 17 and shows evaluation of a query on key range $u$ through $y$. First, note that the partition filter shows that original partitions 1 and 4 do not have any records in that key range. E.g., given a query selecting range $u - y$ from the data shown in Figure 18, the bitmap would indicate that only partitions #2, #3, and #5 need to be merged. As in Figure 17, all remaining records were moved out of partition 5 into partition 6, and partition 5 was eliminated. Note that in partitioned B-trees, a partition is eliminated simply by deletion of records. Second, notice that the data in partition #6 (records from $h - l$) are not read. Therefore, the results of the merge step (which fall into the range $t - z$) can be simply appended to the existing records in partition 6. In fact, leaf nodes in the final merge partition may be accessed write-only except for those that directly overlap with merge results.

Figure 20 sketches the processing of a fifth query, on a point query, searching for records with a key value of $t$. In this case, when updating the final merge partition, only the records that fall into the range $t-z$ need be merged with those found from partitions $1 - 4$.

Finally, note that in the figure, the partition filters' bitmap is updated to reflect the addition of the new records to the final merge partition, but that the updated bitmap (shown in the lower half of the figure) does not reflect the movement of records out of partitions. Delaying updates of the partition filters

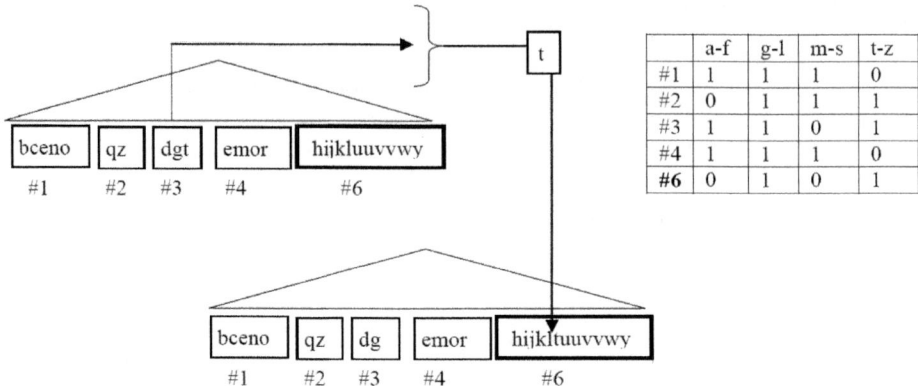

**Fig. 20.** Partition filters make point queries particularly efficient

means that means that partition filters can definitely indicate the absence of a particular key value from a partition, but not its presence. Delayed propagation of deletions to partition filters is just one technique. Other options include immediately propagating deletions, or periodically recalculating partition filter information.

### 5.3 Summary

The primary difference is heap versus B-tree, with the former achieving whatever performance it might immediately after loading and the latter improving its performance as side-effect of query execution. Zones and zone filters are only as effective as the degree to which the workload corresponds to the filter information. For example, if queries in the workload focus upon a particular range of values for a key column, then both the partition filter and zone filters would be able to guide queries to search just relevant collections of data. If, on the other hand, the queried column is not a key one (so data is completely uniformly distributed across all zones), then all zones would need to be scanned for each query instance. In contrast, adaptive merging would merge the queried records into a single partition upon the first query.

Furthermore, because searching an unordered heap requires a full scan, zones are limited to sizes that can be quickly scanned, such as the 3 MB of disk space in Netezza's implementation or the 64 K records in the Brighthouse design. In contrast, partition filters apply to partitions in a B-tree, which allows much larger quantities of data to be searched much more efficiently.

Overhead at query time represents the cost of performing the incremental optimization of the index and are due not to false hits or to partitions being too large, but rather depend on the current state of the index with regard to the queried range. Because under adaptive merging the index starts off only partially organized and then is improved as a side-effect of queries, different queries will incur different overheads. Given a query with a predicate on a specific range, one of the following situations will occur. (1) If the queried range is completely contained in a fully-optimized region of the index (the final merge partition), and there will be no overhead over the cost of performing the query in the presence of a full index. Alternatively, (2) if the queried range does not fall within the final merge partition, then the overhead will reflect the cost of merging query results into the fully-optimized final merge partition. Insofar as (2) reflects the merging of partitions that don't contain query results, then partition filters may reduce the number of partitions that must be merged at runtime.

Again, we do not consider policy decisions such as index placement (e.g., in memory, on flash, or on traditional disk), and we consider neither removal of indexes (e.g., due to frequent updates) nor heuristic guidance that triggers, delays, or prevents creation of specific indexes. Our focus at this time is on mechanism, not policy on the data structures and the algorithms, not on their placement in the memory hierarchy or in an overall system.

The proposed techniques should be tuned with regard to the memory size of the database server. In particular, the following two parameter settings should reflect allocated memory: (1) the size of initial adaptive merging partitions and (2) the number of partitions merged at a time into the final merge partition. If there are more partitions than can comfortably be merged inmemory, then multiple levels of merging may be needed, performing one merge level per query. But a single query will not incur the entire cost of executing a multi-level merge.

## 6   Performance

In order to assess the value of zone filters and zone indexes, we implemented an approximation using a commercial database product and the "line item" table of the well-known TPC-H database. The following three subsections demonstrate the value of zone maps with low and high values, of zone filters with bit vector filters, and of zone indexes. Instead of building multiple non-clustered indexes, we created only a clustered index on two of the three date columns in each table. Given that this particular database has no outlier values, we kept the number of low and high values to one each in the experiments ($m = n = 1$). Instead of bit vector filters, we used small tables.

The table contains about 60 M rows ( 6 GB). It is stored in a clustered B-tree index organized on "commit date," with 2,466 distinct values and about 2 MB of data per distinct commit date. The "ship date" is a secondary sort column in the clustered index. For the "receipt date," there is only an incidental order due to correlation; this correlation is not exploited during query processing.

The hardware is a dual-core Intel T7200 CPU running at 2 GHz, 2 GB of RAM, a SATA drive with all system files, and a PATA drive with the test database (and its recovery log, but nothing else). Both drives use NTFS file systems recently defragmented. The database server is limited to 128 MB of workspace in RAM. While the system ran other applications concurrently, the CPU utilization was very low throughout all experiments and the times reported below reflect I/O times very accurately. Each statement started with a warm procedure cache (queries compiled ahead of time) and a cold buffer pool.

In all following experiments, we focused on measuring the overhead incurred when index structures are modified adaptively as a side-effect of the read-only queries. We did not address the overhead incurred when maintaining the adaptive index in response to updates to data values. The experimental workloads are simple queries that aim to test performance across a range of parameter values. We did not focus on traditional "benchmark" workloads such as TPC-PH or TPC-DS because we wanted to isolate the benefits and overhead of each of our proposed improvements and track its progress across the lifetime of the workload.

## 6.1   Zone Filters

Zone filters are a combination as well as a generalization of small materialized aggregates, zone maps, and the bit vector filtering techniques in the Brighthouse and Kognitio products. The following experiments do not explore and demonstrate their full power; nonetheless, they enable an assessment and appreciation of the performance effects compared to a traditional full table scan.

## 6.2   Zone Maps

In order to emulate a zone filter with lowest and highest values per column, the database also contains a materialized view that captures lowest and highest receipt date for each ship date. These columns have high correlation, but the ranges of receipt dates overlap for a number of neighboring commit dates. In other words, this is precisely the constellation for which Netezza's zone maps are designed, with the ordering on commit dates emulating a load sequence in a Netezza database.

Table 1 shows relevant SQL text and the observed performance. All queries in the experiments are variation on the first query shown. If this query can exploit a clustered index, it merely scans a contiguous key range and counts 374,382 rows at about 100,000 rows per second (10 MB/s). The first variation of this query forces a complete table scan. Scanning 6 GB in 682 seconds also indicates about 9 MB per second. The performance difference between a disk-order scan and an index-order scan is probably not significant as the database was recently defragmented.

Creation of a zone map employs a very simple plan here, effectively a complete scan of the clustered index with very moderate effort for aggregation calculations plus insertion into a new table serving as zone map in our emulation of this technology. The small performance difference between the prior query and the creation of a zone map indicates that creation of a zone as side effect of loading is a good design.

The final query in Table 1 shows the effect of zone maps. (The final line ensures the desired query execution plan.) Using the emulated zone map, the query completes 10 times faster than the equivalent query without a zone map. This closely mirrors the fact that about 9

Creation of zone filters with m+n lowest and highest values takes the same effort as creation of zone maps, as discussed above. In a table without Null

**Table 1.** Effect of zone maps

| Experiment | SQL Text | |
|---|---|---|
| Query using a clustered index 3.75 sec | Select | count (*) |
| | From | lineitem |
| | Where | l_commitdate between '06/16/1994' and '06/30/1994' |
| Query without a useful index 682 sec | Select | count (*) |
| | From | lineitem |
| | Where | l_receiptdate between '06/16/1994' and '06/30/1994' |
| **Zone map creation** **691 sec** | Select | l_commitdate as zone, min (l_receiptdate) as low, max (l_receiptdate) as high |
| | Into | map_receiptdate |
| | From | lineitem |
| | Group by l_commitdate | |
| **Query using a zone map** **54.8 sec** | Select | count (*) |
| | From | map_receiptdate, lineitem |
| | Where | low <= '6/30/1994' and high >= '6/16/1994' and l_commitdate = zone and l_receiptdate between '6/16/1994' and '6/30/1994' |
| | option (loop join) | |

values and without outlier values, like the TPC-H table in this experiment, the performance effect of zone filters also equals that of zone maps. If Null values and outliers are present, however, zone maps are likely to be useless, such that query processing resorts to an unguided table scan. Zone filters with m+n lowest and highest values, on the other hand, are much more likely to let scans skip over many database segments of zones, such that query performance is like the last case in Table 1.

### 6.3  Bit Vector Filtering

In order to emulate bit vector filtering in a zone filter, a materialized view captures the distinct part numbers for each commit date. The performance difference between query execution plans that do or do not use this materialized view indicates the value of bit vector filtering in zone filters.

Table 2 shows SQL statements for creation and usage of emulated zone filters. 64,000 bits might seem relatively large compared to a zone map with only minimum and maximum value, but 64,000 bits or 8 KB are a fairly dense synopsis for a database segment or zone of 2 MB and a domain of 2,000,000 values.

In the emulation, however, due to the lack of an appropriate bitmap data type, individual records are created such that the materialized view contains 50,000,000 records or about 2,000 records per distinct value in the commit date

**Table 2.** Effect of zone filters

| Experiment | SQL Text | |
|---|---|---|
| **Zone filter creation 1,656 sec** | Select | distinct l_commitdate as zone, checksum (l_partkey) % 64000 as value |
| | Into | filter_partkey |
| | From | lineitem |
| **Query using a zone filter 300 sec** | Select | count (*) |
| | From | (select distinct zone from filter_partkey where value = checksum (1705409) % 64000) as filter, lineitem |
| | Where | l_commitdate = zone and l_partkey = 1705409 |
| | option (loop join) | |

column. The size of the table is about 1 GB. With the experimental system, a scan of 1 GB requires about 100 seconds. Nonetheless, even this poor zone filter eliminates a large number of database segments from the scan of the line item table, sufficient for a faster query execution time than a table scan without indexes and zone filters.

With proper bitmaps, the size of all zone filters would be about 2,466  8 KB = 20 MB and the set of all zone filters could be scanned in 2 seconds. The observed query time would not be 300 seconds but about 200 seconds. Without bit vector filtering, all database segments or zones need to be scanned, and the query execution time equals that of the query without a useful index in Table 1. In other words, the benefit of zone filters with bit vector filters can be substantial, even where traditional zone maps may completely fail.

## 6.4   Zone Indexes

In order to assess the value of zone indexes, another variation of the base query of Table 1 restricts the ship date, as shown in Table 3. This query is very similar to the last query of Table 1. Both queries exploit zone maps to skip over irrelevant database segments or zones. The difference is that a restriction on the receipt date in Table 1 must scan each relevant database segment in its entirety, whereas a restriction on the ship date can exploit the secondary index order. This secondary index order is basically equivalent to having an index on ship date within each database segment.

The performance difference demonstrates the value of an index within each database segment, with 55.8 seconds versus 7.95 seconds. In fact, the elapsed time of the query in Table 3 is remarkable not only for the performance difference relative to the last query of Table 1 but also for the performance similarity

**Table 3.** Effect of zone indexes

| Experiment | SQL Text |
|---|---|
| **Query using a zone index 7.95 sec** | Select     count (*) <br> From      map_shipdate. lineitem <br> Where     low <= '6/30/1994' <br>              and high >= '6/16/1994' <br>              and l_commitdate = zone <br>              and l_shipdate between <br>              '6/16/1994' and '6/30/1994' <br> option (loop join) |

relative to the first query of Table 1. Their difference is merely a factor of 2 in spite of the fact that creation and maintenance of a clustered index requires effort and indeed reduces load bandwidth to a fraction of the hardware bandwidth. Zone filters and zone indexes, on the other hand, can easily be created as side effects as the unsorted load stream passes through memory.

## 6.5   Emulating Partitioned B-Trees

The following experiments use an alternative physical database design for the lineitem table, with the clustered index organized as a partitioned B-tree index. With a product that natively supports partitioned B-tree indexes, values in the artificial leading key field are assigned entirely automatically during loading and the implicit run generation during loading. Emulating a load sequence based on order date, and assuming a fixed interval between order date and commit date, the artificial leading key field in the partitioned B-tree is equal to the week of the commit date (weeks since 1992-01-01). Thus, the initial B-tree contains 353 distinct values in the artificial leading key field. 353 partitions imply an average size of 60 GB   353 = 170 MB per partition. Assuming run generation using replacement selection, this corresponds to a workspace of about 85 MB, which seems realistic for index creation on modern machines.

Table 4 shows some fragments of SQL text and the elapsed times on our machine. The first statement copies the lineitem table into a new table, extended

**Table 4.** Partitions within a clustered index

| Elapsed time (mm:ss) | SQL text |
|---|---|
| 13:15 | select ('1992-01-01' − l_commitdate) / 7 as commitweek , <br>          ('1992-01-01' − l_receiptdate) / 7 as receiptweek , <br>          lineitem.* <br> into aux0 <br> from lineitem |
| 48:02 | create clustered index partd on aux0 (commitweek) |
| 6:27 | select count (*) from aux0 where receiptweek = 160 |

with two columns counting weeks from the earliest date in the table. The new table called "aux0" is the main table in the subsequent experiments. The second statement creates a clustered index that emulates a partitioned B-tree created during loading. As shown in the table, it takes 48:02 to create a complete index.

The third statement establishes the baseline performance for query execution with partitioned B-trees, adaptive merging, and partition filters. As shown in the table, it takes 6:27 to perform a full table scan.

Comparing the times required for a full table scan and for index creation illustrates how traditional index creation can interfere with query processing. This difference demonstrates the need for alternatives to explicit index creation and expensive index maintenance during data import.

## 6.6   Partition Filters

The next set of experiments measures the effect of partition filters in terms of the overhead of their creation and their value during query processing. The experiment measures partition filters applied to a partitioned B-tree; their effect would be quite similar for partitioning in other contexts, like heaps or zones.

**Table 5.** Partition filters

| Elapsed time (mm:ss) | SQL text |
|---|---|
| 6:29 | create table aux1<br>     (commitweek int primary key,<br>     lowreceiptweek datetime,<br>     highreceiptweek datetime)<br>insert into aux1 (commitweek, lowreceiptweek, highreceiptweek)<br>     select  commitweek, min (receiptweek), max (receiptweek)<br>     from   aux0<br>     group by commitweek |
| 6:34 | select count (*) from aux0 where receiptweek = 160 and<br>     commitweek in (select commitweek from aux1) |
| 0:35 | select count (*) from aux0 where receiptweek = 160 and<br>     commitweek in (select commitweek from aux1 where<br>          160 between lowreceiptweek and highreceiptweek) |

The first statement (pair) in Table 5 creates and populates a simple partition filter for the new table. This pair of SQL statements emulates the creation of a partition filter. The small difference between query execution (see Table 4) and filter creation indicates how small overhead of filter creation during loading.

The second statement in Table 5 shows the case in which a partition filter is inspected but has no value towards reducing the scan effort in the main table. In other words, it measures the worst case for query processing with partition filters. The query execution plan chosen iterates over the rows in the partition filter and, for each row, scans the matching partition in the main table. The overhead in the worst case is small, only a few seconds of elapsed time. Worst case is you lose a few seconds.

The third statement in Table 5 illustrates the advantage gained from partition filters. 32 of 353 partitions contain rows satisfying the query predicate, and query execution time is indeed less than 10

## 6.7   Adaptive Merging with Partition Filters

The final set of experiments illustrates cost and benefit of adaptive merging with partition filters.

**Table 6.** Adaptive merging

| Elapsed time (mm:ss) | SQL text |
|---|---|
| 6:50 | update aux0 set commitweek = 999 where receiptweek = 162 |
| 2:44 | update aux0 set commitweek = 999 where receiptweek = 160 and commitweek in (select commitweek from aux1 where 160 between lowreceiptweek and highreceiptweek) |
| 0:02 | select count (*) from aux0 where receiptweek = 160 and commitweek in (999) |

Table 6 shows the overhead of reorganization during query execution. The first statement in Table 6 shows the complete cost, including the overhead of incremental index construction, for the second query in a workload before the index has been constructed. The execution time for this query, 6:50, includes the full overhead incurred by adaptive merging when not using partition filters. The second statement in Table 6 uses an ordinary SQL update command to achieve the effect of adaptive merging with partition filters. This query's execution time demonstrates how partition filters reduces the overhead of the incremental update cost from 6:50 to 2:44. Processing these updates as side effects of query execution might enable some additional efficiencies, e.g., exploiting the sort order of insertions and deletions as well as not logging record contents. Compared to the cost of merely scanning records in multiple places, moving those records to a new destination partition more than quadruples the cost. This is because a pure query requires only one I/O per page; the poor emulation using a commercial database also writes database pages and log pages for both deletions and insertions.

The third statement in Table 6 demonstrates the advantages to be gained by adaptive merging. In fact, query performance equals that of query processing using a traditional B-tree index created a priori at a substantial cost and maintained throughout all load operations. The difference of query processing performance in a partitioned B-tree left behind by full-bandwidth loading and in a fully optimized B-tree is so large that less than half a dozen queries for a given key range suffice to outweigh the cost of B-tree optimization even in this poor emulation of adaptive merging. If adaptive merging is deeply integrated into a database system, the break-even point would be reached with mere 3 queries for any given key range in the index.

Taken together, partitioned B-trees enable bandwidths during initial and incremental load operations comparable to the hardware write bandwidth; partition filters alleviate the cost of searching for query results in a highly partitioned table or index; and adaptive merging ensures that query performance improves automatically and incrementally until it equals that of query processing in a traditional database indexed a priori at great expense and with perfect (and thus impossible) foresight.

## 7   Summary and Conclusions

In summary, we identified three advantageous characteristics and three limiting characteristics in Moerkotte's "small materialized aggregates" and Netezza's "zone maps." The advantages are that data can be loaded directly into heaps, that there is no need for indexes or index tuning, and that scans for typical queries are very fast. The disadvantages are that query predicates on columns not correlated with the load sequence may require scanning all heaps, that search within a heap requires a scan, and that a single outlier value can cause the scan of an entire heap. The generalizations of those prior designs proposed in this article overcome their limitations yet retain their advantages.

First, zone filters generalize from minimum and maximum (for each column and each database segment) to the m lowest and n highest values plus a bit vector filter. The former enables the filter to work effectively even in the presence of Null values and outliers in the data. The latter enables the filter to work effectively even for columns without correlation to the load sequence. Zone filters, like zone maps, are sufficiently inexpensive in creation and storage that they can be applied to all columns in a table.

Second, partition filters generalize the concept from zones defined by storage to partitions defined by data values. Partitions apply to nodes in shared-nothing systems, partitions within a single node, partitions in a partitioned B-tree and its artificial leading key field, or any other form of partitions.

Third, zone indexes enable efficient search within a database segment. They can be created very efficiently during database loading and refined during query processing, without external storage or I/O. During query processing, they enable very efficient search within a large segment. Complete indexing is possible by storing each value only once, very much like in-memory indexes such as T-trees. Thus, there is no need for traditional index tuning and no risk of mistaken omission of valuable indexes.

Together, these techniques enable efficient query processing immediately after a high-performance load. These structures can be created as a side effect of load processing, with moderate memory and processing needs. No intermediate database reorganization is required in order to optimize raw data left behind by the load operation. Query processing eliminates many database segments in most cases and can search the remaining ones very efficiently.

In addition to loading traditional data warehouses, the technique might prove useful in capturing data streams for future auditing as well as immediate query

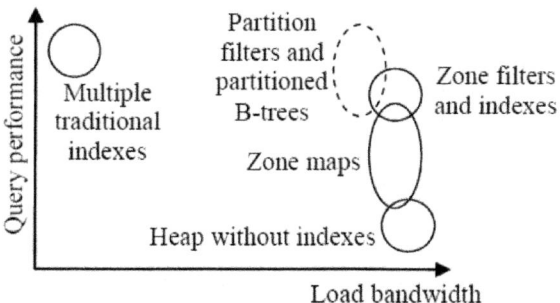

**Fig. 21.** Alternative techniques at a glance

processing, e.g., matching data from multiple streams. The effort for sorting database segments in memory is similar to creation of in-memory indexes, yet zone filters and zone indexes are equally suitable for disk storage and thus for very large data streams.

Figure 21 illustrates the tradeoffs among the techniques. Multiple traditional indexes permit optimal query performance at the expense of rather poor load performance. Heaps without indexes permit load at hardware write bandwidth but lead to poor query performance, in particular for highly selective queries. Partitioned B-trees combine high load bandwidth and high query performance but only at the expense of intermediate operations that reorganize and optimize the B-tree indexes. Thus, the load bandwidth that can be sustained over a long period is less than the ideal load bandwidth by a small factor. If one index and its partitions are used to implement deferred updates for other indexes, load performance can approach hardware write bandwidth. Zone maps permit the load bandwidth of heaps and enable good query performance but only in cases of correlation. Finally, a table with zone filters and zone indexes can be loaded at hardware write bandwidth, requires no reorganization or optimization after loading, and instantly can enable query performance comparable to multiple indexes as shown in the experiments.

In conclusion, we believe that this design is another step towards resolving the tension between effort during database loading and during query execution. In a traditional database implementation, a user must choose between index maintenance during load processing, index creation after load completion and recreation after each incremental load or large scans during query execution even for simple predicates amenable to efficient index support. Partitioned B-trees enable load operations to append new records in multiple indexes but require merging of partitions after load completion. Zone maps can be created efficiently during load processing and guide large scans to skip over needless database segments, but they are effective only for queries and predicates that restrict a column with a high correlation with the load sequence. Zone filters and zone indexes permit efficient load operations, apply to many more queries and predicates, and guide the search within a large database segment in order to minimize faults in the CPU caches and time spent on predicate evaluation.

In the future, we plan on investigating the most appropriate sizes of database zones depending on the ability to exploit multiple predicates in a zone filter, on the internal organization using zone indexes, and on the storage medium, in particular traditional disks and flash memory devices. We believe that the partitioned B-trees and the adaptive merging approaches work better with block access devices such as both flash memory and magnetic disks better than other techniques for adaptive indexing. [G 07b]. We suspect that a crucial parameter is the device-specific product of access latency and transfer bandwidth, i.e., the I/O size for which access latency equals transfer time. For the wide variety of optimal zone sizes we expect, we plan on investigating the optimal data organization within each zone, e.g., NSM, PAX, or a hybrid of these two formats [ADH 01, G 08]. We suspect that a column-optimized format will emerge with two sizes, namely small pages optimized for flash memory within large pages optimized for traditional disks, with the former also defining the mini-pages of PAX [ADH 01]. An additional challenge with this data organization is the optimization of zone indexes in single and multiple columns, including appropriate algorithms for search [LJB 95] and for sorted retrieval. Finally, a recent collaboration focuses on adaptive indexing techniques, considering database cracking and adaptive merging in the context of both in-memory and disk-based databases, and includes extending both systems to handle updates [IMGK 10].

## Acknowledgements

Stratos Idreos and Stefan Manegold have very graciously and generously contributed insightful comments and suggestions to our papers on adaptive merging.

## References

[ADH 01]   Ailamaki, A., De Witt, D.J., Hill, M.D.: Marios Sk- ounakis: Weaving relations for cache performance. In: VLDB 2001, pp. 169–180 (2001)
[B 70]     Bloom, B.H.: Space time trade-offs in hash coding with allowable errors. ACM Commun. 13(7), 422–426 (1970)
[BD 83]    Boral, H., DeWitt, D.J.: Database machines: An idea whose time passed? In: A critique of the future of database machines. In: IWDM 1983, pp. 166–187 (1983)
[BU 77]    Bayer, R., Unterauer, K.: Prefix B-trees. ACM TODS 2(1), 11–26 (1977)
[CDC 08]   Cao, Y., Das, G.C., Chan, C.Y., Tan, K.-L.: Optimizing complex queries with multiple relation instances. In: SIGMOD 2008, pp. 525–538 (2008)
[F 94]     Fernandez, P.M.: Red Brick Warehouse: A read-mostly RDBMS for Open SMP platforms. In: SIGMOD 1994, p. 492 (1994)
[G 03]     Graefe, G.: Sorting and indexing with partitioned B-trees. In: CIDR 2003 (2003)
[G 07a]    Graefe, G.: Hierarchical locking in B-tree indexes. In: BTW 2007, pp. 18–42 (2007)

[G 07b]     Graefe, G.: The five-minute rule twenty years later, and how flash memory
            changes the rules. In: DaMoN 2007, p. 6 (2007)
[G 08]      Graefe, G.: Integrating PAX and NSM page formats. Hewlett-Packard Lab-
            oratories (2008) (unpublished manuscript)
[GK 10a]    Graefe, G., Kuno, H.: Self-selecting, self-tuning, incrementally optimized
            indexes. To appear in EDBT (2010)
[GK 10b]    Graefe, G., Kuno, H.: Two adaptive indexing techniques: improvements
            and performance evaluation (submitted)
[GK 10c]    Graefe, G., Kuno, H.: Adaptive indexing for relational keys. To appear in
            SMDB (2010)
[GKK 01]    Grtner, A., Kemper, A., Kossmann, D., Zeller, B.: Efficient bulk deletes in
            relational databases. In: ICDE 2001, pp. 183–192 (2001)
[GL 01]     Graefe, G., Larson, P.-k.: B-Tree indexes and CPU caches. In: ICDE 2001,
            pp. 349–358 (2001)
[IKM 07a]   Idreos, S., Kersten, M.L., Manegold, S.: Database cracking. In: CIDR 2007,
            pp. 68–78 (2007)
[IKM 07b]   Idreos, S., Kersten, M.L., Manegold, S.: Updating a cracked database. In:
            SIGMOD 2007, pp. 413–424 (2007)
[IKM 09]    Idreos, S., Kersten, M., Manegold, S.: Self-organizing tu-ple reconstruction
            in column stores. In: SIGMOD 2009, pp. 297–308 (2009)
[IMGK 10]   Idreos, S., Manegold, S., Graefe, G., Kuno, H.: Adaptive indexing. Sub-
            mitted for publication (2010)
[JDO 99]    Jermaine, C., Datta, A., Omiecinski, E.: A novel index supporting high
            volume data warehouse insertion. In: VLDB 1999, pp. 235–246 (1999)
[KM 05]     Kersten, M.L., Manegold, S.: Cracking the database store. In: CIDR 2005
            (2005)
[L 01]      Lomet, D.B.: The evolution of effective B-tree page organization and tech-
            niques: a personal account. SIGMOD Record 30(3), 64–69 (2001)
[LBM 07]    Lang, C.A., Bhattacharjee, B., Malkemus, T., Wong, K.: Increasing buffer-
            locality for multiple index based scans through intelligent placement and
            index scan speed control. In: VLDB 2007, pp. 1298–1309 (2007)
[LC 86]     Lehman, T.J., Carey, M.J.: A study of index structures for main memory
            database management systems. In: VLDB 1986, pp. 294–303 (1986)
[LJB 95]    Leslie, H., Jain, R., Birdsall, D., Yaghmai, H.: Efficient search of multi-
            dimensional B-trees. In: VLDB 1995, pp. 710–719 (1995)
[M 08]      Monash, C.: Kognito WX2 overview (January 2008),
            http://www.dbms2.com/2008/01/26/kognitio-wx2
[M 98]      Moerkotte, G.: Small materialized aggregates: A light weight index struc-
            ture for data warehousing. In: VLDB 1998, pp. 476–487 (1998)
[MKY 81]    Merrett, T.H., Kambayashi, Y., Yasuura, H.: Scheduling of Page- Fetches
            in Join Operations. In: VLDB 1981, pp.488–498 (1981)
[MOP 00]    Muth, P., O'Neil, P., Pick, A., Weikum, G.: The LHAM log-structured
            history data access method. VLDB J. 8(3-4), 199–221 (2000)
[MR 93]     Murphy, M.C., Rotem, D.: Multiprocessor Join Scheduling. IEEE
            TKDE 5(2), 322–338 (1993)
[NBC 95]    Nyberg, C., Barclay, T., Cvetanovic, Z., Gray, J., Lomet, D.B.: AlphaSort:
            A Cache-Sensitive Parallel External Sort. VLDB J. 4(4), 603–627 (1995)
[RR 00]     Rao, J., Ross, K.A.: Making B+-trees cache conscious in main memory. In:
            SIGMOD 2000, pp. 475–486 (2000)

[SWE 08]   Slezak, D., Wroblewski, J., Eastwood, V., Synak, P.: Brighthouse: an ana-
           lytic data warehouse for ad-hoc queries. PVLDB 1(2), 1337–1345 (2008)
[V 87]     Valduriez, P.: Join Indices. ACM TODS 12(2), 218–246 (1987)
[ZHN 07]   Zukowski, M., Hman, S., Nes, N., Boncz, P.A.: Cooperative scans: Dynamic
           bandwidth sharing in a DBMS. In: VLDB 2007, pp. 723–734 (2007)
[ZLF 07]   Zhou, J., Larson, P.-k., Freytag, J.C.: Wolfgang Lehner: Efficient exploita-
           tion of similar subexpressions for query processing. In: SIGMOD 2007, pp.
           533–544 (2007)

# Efficient Online Aggregates in Dense-Region-Based Data Cube Representations*

Kais Haddadin[1] and Tobias Lauer[2]

[1] Jedox AG, Freiburg, Germany
kais.haddadin@jedox.com
[2] Institute of Computer Science, University of Freiburg, Germany
lauer@informatik.uni-freiburg.de

**Abstract.** In-memory OLAP systems require a space-efficient representation of sparse data cubes in order to accommodate large data sets. On the other hand, many efficient online aggregation techniques, such as prefix sums, are built on dense array-based representations. These are often not applicable to real-world data due to the size of the arrays which usually cannot be compressed well, as most sparsity is removed during pre-processing. A possible solution is to identify dense regions in a sparse cube and only represent those using arrays, while storing sparse data separately, e.g. in a spatial index structure. Previous dense-region-based approaches have concentrated mainly on the effectiveness of the dense-region detection (i.e. on the space-efficiency of the result). However, especially in higher-dimensional cubes, data is usually more cluttered, resulting in a potentially large number of small dense regions, which negatively affects query performance on such a structure. In this article, our focus is not only on space-efficiency but also on time-efficiency, both for the initial dense-region extraction and for queries carried out in the resulting hybrid data structure. After describing a pre-aggregation method for representing dense sub-cubes which supports efficient online aggregate queries as well as cell updates, our sub-cube extraction approach is outlined in detail. In addition, optimizations in our approach significantly reduce the time to build the initial data structure compared to former systems. Two methods to trade available memory for increased aggregate query performance are provided. Also, we present a straightforward adaptation of our approach to support multi-core or multi-processor architectures, which can further enhance query performance. Experiments with different real-world data sets show how various parameter settings can be used to adjust the efficiency and effectiveness of our algorithms.

## 1 Introduction

Online analytic processing (OLAP) allows users to view aggregate data from a data warehouse displayed on demand, using a model that is usually referred to as the *data cube* [6], which includes operations such as slicing and dicing as well as roll-up and drill-down along hierarchies defined over dimensional attributes. Depending on the

---

* This article is an extended version of [9].

A. Hameurlain et al. (Eds.): TLDKS II, LNCS 6380, pp. 73–102, 2010.
© Springer-Verlag Berlin Heidelberg 2010

architecture of the OLAP system, the aggregate values are either pre-computed and stored or calculated from the base data on the fly, i.e. only when the respective value is requested. (Combinations of both approaches are also very common.) The latter strategy may result in longer times for retrieving and aggregating the necessary base values – especially if they have to be loaded from secondary storage media – but it is usually faster regarding the changes of cell values, as expensive re-computation of stored aggregates is avoided. In addition, this approach is used by most in-memory OLAP databases which try to store all of the base data in RAM and calculate all aggregate values online (except for recently computed aggregates residing in a cache).

Efficient online aggregation can be achieved by transforming the base data according to some pre-processing strategy. Well-known examples are the prefix-sum approach [10] and its variants, which allow the computation of arbitrary range queries in constant time, usually at the expense of update costs. More sophisticated strategies provide a variety of tradeoffs between query and update times [5, 11, 14]. The iterative data cube (IDC) [14] allows the combination of several of those strategies by choosing a separate one for each dimension.

One common feature of these methods is that that they are based on an (multidimensional) array representation of the data. The advantage is a convenient and efficient access to each cell through its coordinates. However, a serious drawback of the above strategies is that the pre-processing step usually requires filling also the empty cells and hence effectively turns sparse cubes into dense ones. This prevents an efficient compression of the arrays, which thus require much more space than the original sparse data. In particular, this means that data which might otherwise easily fit into main memory cannot be accommodated any more in such a representation.

As a way out of this dilemma, dense-region-based representations have been proposed [2, 3, 13]. These approaches try to exploit what has been called the "dense-region-in-sparse-cube" property [2]: in many real-world datasets, the majority of filled cells are not distributed evenly within the universe of all possible cells but are clustered in certain regions. Each of these regions can be represented as a sub-cube using any of the above methods, while outliers (filled cells that do not belong to any dense region) can be stored separately. Previous approaches have concentrated mainly on effectiveness, i.e. the reduction of sparsity, but have neglected efficiency, both for the pre-processing step of detecting dense regions and for the queries carried out on the resulting data structure.

This article is structured as follows. The following section defines the necessary preliminaries, points to previous work that our approach is related to, and gives an overview of the overall method. Section 3 describes an improved data structure for dense OLAP cubes that is efficient for both aggregate queries along dimensional hierarchies and cell updates in a cube, which can be used to represent each dense sub-cube in our hybrid structure. In section 4, we give a detailed description of our sub-cube extraction procedure with a special emphasis on the time-efficiency of the involved algorithms. Section 5 describes how OLAP range queries are executed on the sparse cube represented by our data structure and also presents a simple way to boost those queries on multi-core or multiprocessor architectures. In section 6, we present detailed experimental results carried out with our approach on test data, before concluding our article and providing an outlook on current and future research.

## 2  Preliminaries and Related Work

For the purposes of this work, we consider a *data cube* $C$ as a $d$-dimensional hyper-rectangle of cells. Each dimension $D_i$ is a discrete range of all $n_i$ possible base values in that dimension; for simplicity, we assume $D_i = \{1, ..., n_i\}$. We call $n_i$ the length or cardinality of dimension $D_i$.

It is obvious that a data cube $C$ consists of $\prod_{i=1}^{d} n_i$ cells. We refer to this number as the *capacity* of $C$. The capacity must be distinguished from the *size* of $C$, which is defined as the number of cells which actually hold a value other than zero. The size thus corresponds to the number of base records stored in a fact table representation of the cube. The *density* of $C$ is defined as the ratio of filled cells to all cells, i.e. $density(C) = size(C) / capacity(C)$. The density of most cubes in real-world OLAP scenarios is very low (usually much less than 1%); this is what is referred to as cube *sparsity*.

Dense-region-based OLAP representations aim at identifying dense sub-cubes within a sparse cube. Each of them can be represented efficiently using the array-based method, as discussed above. They must then be maintained in some index structure allowing quick access to them during queries. An additional data structure is required for storing and accessing the outlier cells, which are not located inside any dense region. Figure 1 gives a rough overview of the data structure used in our implementation of the approach. The basic idea is to extract dense regions and make the resulting sub-cubes and outliers accessible through two instances of the R*-tree, a popular and efficient spatial index structure [1].

Clearly the main goal of any dense-region-based approach is the reduction of sparsity, i.e. the effective identification of clusters of cells. The sparsity reduction can be measured by the *global density* of the resulting set $S$ of sub-cubes, which is defined as

$$density(S) = \frac{\sum_{c \in S} size(c)}{\sum_{c \in S} capacity(c)}$$

Many clustering methods for identifying dense regions have been proposed, especially in the data mining literature, and we refer to 15 for an overview. Since in our scenario, however, the regions will be represented as hyper-rectangles, we aim at identifying rectangular regions and can thus use more straightforward methods.

One of first approaches to dense-region-based OLAP was given in [2]. It proposes extracting dense sub-cubes through an algorithm called *ScanChunk* and then inserting these sub-cubes in an R-tree [8] structure, while outliers are indexed in ROLAP table form. However, the *ScanChunk* algorithm is not efficient for large data cubes. The reason is that its time complexity depends on the capacity, i.e. the number of all possible cells, of the data cube. The capacity of cubes can be extremely large, especially in higher dimensions. Our goal is to find a procedure whose complexity is a function of the *size*, i.e. the number of filled cells, of the data cube rather than its capacity. Also, the clustering result is measured only by the global density. This overlooks another important factor, the number of resulting sub-cubes, which is a decisive factor for query performance.

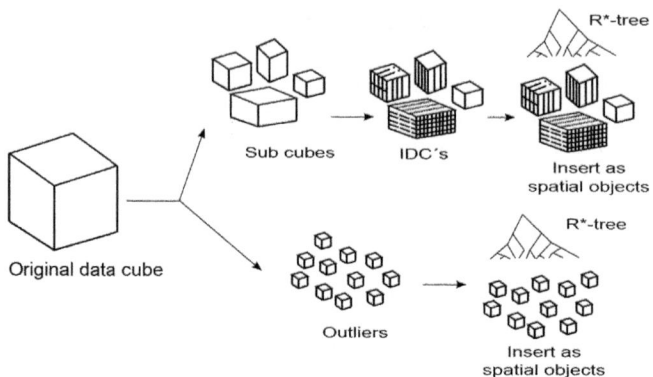

**Fig. 1.** Sketch of the basic approach

Our proposed dense sub-cube extraction procedure is based on the approach proposed in [3] and [13], which divides the procedure into basic steps, splitting, shrinking, merging, and filtering, where the merge step aims at reducing the number of sub-cubes. However, we identified several shortcomings in this step, which are avoided in our approach. In addition, we improve the splitting algorithm which, when implemented as described there, is not time-efficient, especially in cubes of higher dimensions. In addition, the tests described there were done using artificial data of rather low dimensionality ($d \leq 5$), while our own test data was taken from real-world OLAP models with cubes of up to 13 dimensions.

Dense-region-based data structures using tree-like indexing structures also offer themselves for approximate query processing, where internal nodes of the trees hold aggregate information which can be used for fast query result approximation. The *pCube* [15] allows progressive feedback giving users increasingly accurate results over time while providing absolute error bounds. The approach described in [4] delivers approximate query results with probabilistic guarantees. The data structure also maintains outliers separately from non-outlier data, and especially the method of storing and aggregating outliers is similar to ours. However, the strategy is not strictly dense-region-based, as the non-outlier data structure (*TP-Tree*) does not store dense regions but query-driven (sample) partitions of a cube independent of the density. While query approximation is not part of our work, the resulting sets of dense regions and outliers from our refined extraction method could be used in such systems to further improve query response times there.

## 3   A Data Structure for Dense Sub-cubes

In this section, we describe an array-based data structure suitable for representing dense sub-cubes, supporting efficient online aggregate queries as well as updates of cells. It is based on the *space-efficient dynamic data cube* (SDDC) [14], which is a variant of the well-known *prefix-sum* method [10]. We have generalized the approach such that it better supports OLAP hierarchies. Like all the other methods in the tradition of prefix sums, it is space-efficient in the sense that it does not require any more memory than the base data (including zero values), but it is not suitable in itself for sparse cubes.

The prefix-sum method replaces the base values in the cube with prefix sums and thereby achieves constant time for arbitrary range-sum queries at the expense of an $O(n^d)$ update complexity, where $d$ is the number of dimensions and $n$ is the length of each dimension. Several techniques, such as *local prefix sum* or *relative prefix sum* (cf. [5]), have refined this approach, allowing certain tradeoffs between query and update complexity, usually with a constant query time and polynomial time for updates. A notable difference is the SDDC [14], as it achieves sub-linear time for both queries and updates. In the worst case, the number of cell accesses for both range queries and cell updates in a $d$-dimensional SDDC is $O(\log^d n)$. However, while the poly-logarithmic update time is far better than that of the other techniques, the same complexity for queries is not necessarily efficient in practice. In fact, it might be worse than a run-time linear in the number of actually existing records, especially in sparse cubes.

All the above approaches try to improve worst-case times of *arbitrary* range-sum queries. In most real-world applications, however, the overwhelming majority of queries follow certain patterns, such as roll-up, drill-down, slice and dice operations along pre-defined hierarchies. We have therefore modified the SDDC such that aggregate queries along hierarchies are boosted to an average constant number of cell accesses (plus a small logarithmic extra cost), while updates still remain in poly-logarithmic time.

### 3.1 The Space-Efficient Dynamic Data Cube

In an SDDC the base values of a cube are replaced by recursive prefix sums. An example of the pre-computation along one dimension can be seen in Figure 2. The original array $A[0, \ldots, n-1]$ is transformed into an SDDC by replacing the values in the cells below the non-leaf nodes of the (conceptual) tree drawn above $A$. Each tree node contains the sum of all cells below its left subtree plus the node itself. Note that the last cell contains the sum of all cells and hence the root has only a left subtree.

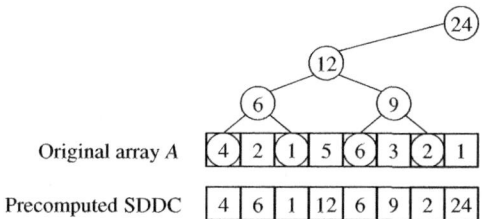

**Fig. 2.** An array $A$ before and after SDDC pre-computation

A prefix-sum $sum[0 : c]$ can be answered easily in an SDDC: descend down the conceptual tree to the node representing $c$ and sum up the values of all those cells on the path that are left of or equal to $c$. In the example in Figure 2, the prefix-sum query $sum[0 : 4]$ adds up the values of cells 3 and 4, resulting in the sum $12 + 6 = 18$, which is the correct value of the sum of cells 0 to 4 in the original array $A$. Clearly, the worst-case (and average-case) cost of such a query is proportional to the height of the implicit tree, which is in $O(\log n)$. An arbitrary one-dimensional range-sum query $sum[l : r]$ (where $l < r$) can obviously be answered by computing the difference of two prefix-sum queries $sum[0 : r] - sum[0 : l\text{-}1]$. Its time complexity is thus also in $O(\log n)$ in the one-dimensional case, and in $O(\log^d n)$ in a $d$-dimensional cube (cf. [14]). Note, however, that several cells may be accessed by *both* of the prefix-sum queries, while they would not have to be accessed at all (since the second query always neutralizes a value which also occurs in the first one). In higher dimensions, this can result in many unnecessary cell accesses. Our modification addresses this issue.

### 3.2 A Generalized Data Structure

In our generalization, we first modify a range-sum query $sum[l : r]$ such that in a pre-processing step we remove duplicate cell accesses before executing the two involved prefix queries $sum[0 : r]$ and $sum[0 : l\text{-}1]$. As we have shown in [11], this modification boosts those range-sum queries of the form $sum[c : c]$ (i.e. the ones which look up a single cell value) to $2^d$ (i.e. a constant number of) cell accesses in the average case. The pre-processing step itself takes $O(d \log n)$ steps, hence dominating the asymptotic query time, which however is still much less than the poly-logarithmic $O(\log^d n)$ of the SDDC.

In order to transfer this improvement to more general range-sum queries, we allow the conceptual tree (cf. Figure 2) to be less perfectly balanced by shaping it in accordance with the hierarchies defined for the corresponding dimension. Figures 3 and 4 show an example of a simple dimension with 19 base elements grouped into four categories of different size; Figure 3 shows the standard SDDC pre-aggregation, (numbers in inner nodes list the range of array positions represented by that node).

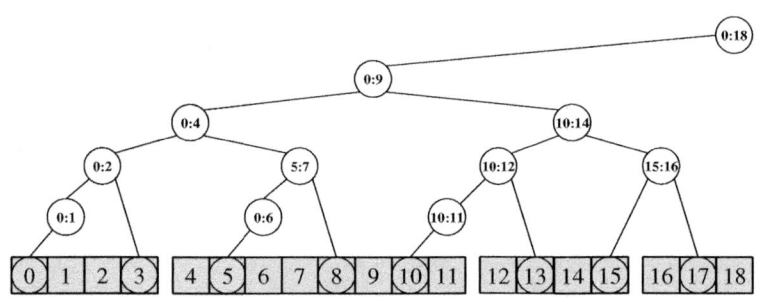

**Fig. 3.** Standard SDDC pre-aggregation for a dimension with 19 elements

A common type of OLAP query would be an aggregation according to the categories. In Figure 3, in order to compute the aggregate value of category 3, we have to calculate $sum[12 : 15] = sum[0 : 15] - sum[0 : 11]$. The first sum involves cells 9, 14 and 15; the second one requires the values in cells 9 and 11. Cell 9 can be eliminated since it occurs in both sums, resulting in three cells to be accessed. Similarly, for category 2 the values of four cells (9 and 11 as well as 2 and 3) must be read. Categories 1 and 4 require two and four cell accesses, respectively. On average we need 3.25 cells per category.

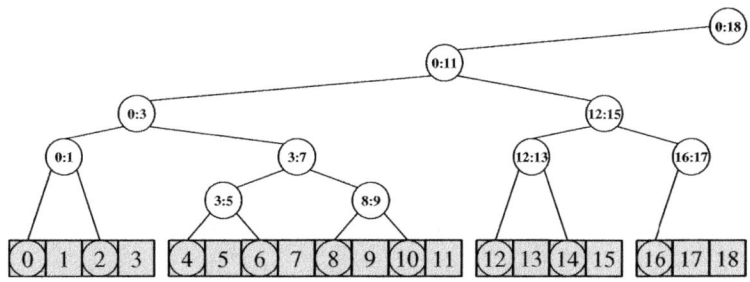

**Fig. 4.** Pre-aggregated dimension using alternative split positions

To improve this, we can adapt the pre-computation by splitting the ranges at the boundaries given by the hierarchy. Figure 4 shows the same dimension pre-aggregated such that the initial split positions (represented by the nodes on the top 3 levels of the tree) coincide with right endpoints of the categories. While this tree may not be as well-balanced as the original one, the aggregate values for the categories can now be established accessing a much smaller number of cells (1, 2, 1 and 3, respectively, i.e. only 1.75 on average). In fact, we have shown in [11] that no more than two cells have to be accessed on average. An average-case analysis shows that queries along any elements of the hierarchy enjoy the same $O(2^d)$ bound for the number of cell accesses as those for base elements.

Note that the flexibility introduced by the modification can result in unbalanced trees, which – in theory – might negatively affect update times. However, experimental results have shown that update times remain comparable with the standard SDDC, while query times are significantly boosted [11].

## 4   Efficient and Effective Sub-cube Extraction

In this section, we describe the individual steps of the extraction procedure, which divides the original cube into a set of dense sub-cubes and a set of outliers. The basic procedure is similar to the ones described in [3] and [13]; it consists of first splitting the cube recursively into dense and sparse regions, then shrinking the dense areas if their borders are too sparse, merging nearby dense regions (to reduce their number), and finally filtering out very small dense sub-cubes (also to reduce the number), treating the filled cells as outliers.

It improves the above approaches in that it achieves the extraction more efficiently and allows for the adjustment of the outcome (in particular, the size of the resulting set of sub-cubes) in order to enhance query performance.

### 4.1   Recursive Splitting

We adopt the basic splitting method from [13] where for each dimension of the cube a histogram is computed which divides the dimension in dense and sparse intervals. The histogram $h_i$ of $D_i$ is an array of length $n_i$, where the value $h_i[j]$ is the number of filled cells in the $j$-th slice of the cube when sliced along $D_i$. The histograms are then used

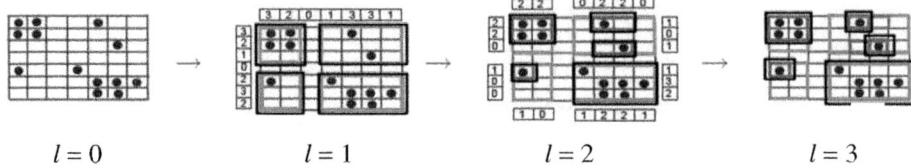

$l = 0$      $l = 1$      $l = 2$      $l = 3$

**Fig. 5.** Recursive splitting of a cube into dense sub-cubes

to detect dense intervals in each dimension by comparing the values against a predefined threshold $\alpha$. The dense intervals then will only include cells with value greater than or equal to $\alpha$. In Figure 5, using $\alpha = 1$ produces two dense intervals in each dimension of the initial cube. The histograms are depicted as the arrays next to the cube in the two center stages.

The Cartesian product of all dense intervals over all dimensions creates a set of sub-cubes, which are candidates for dense regions. The extraction procedure in [13] creates all these sub-cubes and then invokes the same splitting procedure recursively for them. Apparently, many of these candidate cubes will be empty, and computing the histograms for them is neither necessary nor useful. Our initial tests found that even checking them for emptiness to filter them out takes a lot of time. Moreover, the method suggested for computing the histograms via a Boolean array in [3] is not very efficient either. First, the array has a size proportional to the capacity of cube $C$ (rather than its size) and hence should be compressed in order to be handled. Second, in order to compute a histogram, all cells (filled or not) of the current cube are checked, which greatly decreases performance. (Recall that the histograms count *filled* cells, hence ideally only they should be looked at.)

Our approach proceeds the opposite way: a coordinate list of only the filled cells for the current cube is created and processed. For each one, we identify the dense interval to which it belongs in each dimension, using the histograms. We then check whether or not that sub-cube has already been created. If yes, we add the cell to the coordinate list of that sub-cube. If not, the cube must be created before. This way, only those cubes are initialized which actually contain any filled cells, which dramatically reduces execution time.

After all existing sub-cubes of the current cube have been created, the same procedure is called recursively for each of them and a variable $l$ storing the current recursion depth is adjusted (cf. Figure 5). As in [13], the recursive splitting stops if a sub-cube is *continuous*, i.e. if all its dimensions consist of only one dense interval.

The main bottleneck in our procedure is the time to check whether a sub-cube already exists, since this involves going over a potentially large list of created cubes. This is especially relevant in the initial call of the recursive procedure, when the histograms contain a large number of dense intervals. To reduce the number of possible sub-cubes, a technique known as *histogram flattening* can be used. Flattening reduces the differences between adjacent histogram cells by averaging the value of each cell with those of its neighbors. The reduction is controlled by a flattening factor $f$, which defines the number of neighbors that is considered on each side. More details about flattening can be found in [13], where it is also proposed. Quite surprisingly, however, it is neither motivated there why flattening should be used, nor does the procedure improve the experimental results there; on the contrary, the sub-cube extraction always deteriorated the quality of the output (i.e. the global density) in the tests.

In our own experiments we found that flattening can greatly reduce the time needed for splitting, especially during the first split of the initial cube. Therefore, in our work, a new factor called the flattening threshold $\mu$ is added to the flattening process. Flattening is applied only to sub-cubes whose depth $l$ in the recursion is less than or equal to $\mu$. So, if $\mu = 0$, there is no flattening at all. For $\mu = 1$, we apply flattening only to the original data cube. If $\mu = 2$, this means that besides the original sub-cube, flattening is applied to the sub-cubes resulting from the first split. Setting $\mu = \infty$ will apply flattening to all sub-cubes on all levels.

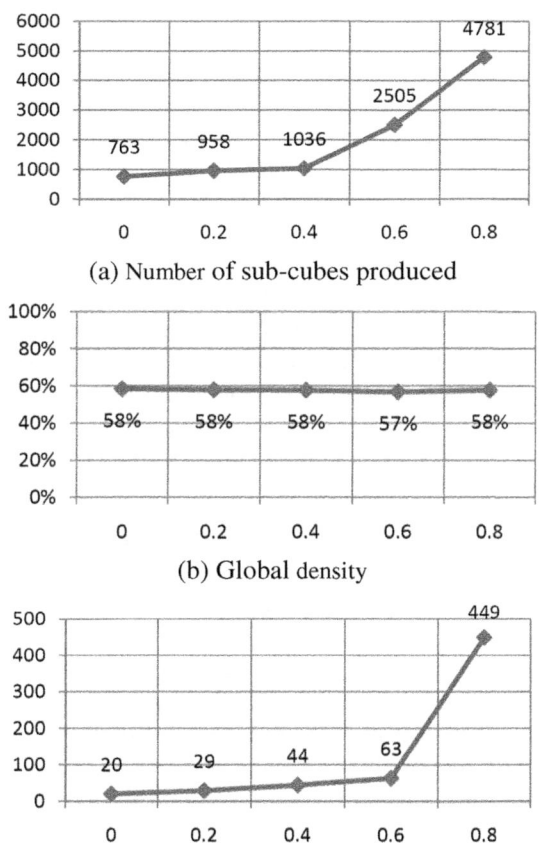

(a) Number of sub-cubes produced

(b) Global density

(c) Time required for remaining extraction procedure (in sec).

**Fig. 6.** Effects of shrinking for different values of $\beta$ (x-axis)

The effects of flattening vary for different values of $\mu$. The speedup in splitting is very big when flattening is applied to the first level of splitting ($\mu = 1$), but gets smaller as the splitting is called to further levels of the split procedure. The reason is that the number of sub-cubes produced at a deeper level of splitting will not be split into many new ones. On the other hand, applying flattening only to the sub-cubes at the first levels of splitting will not reduce the final number of sub-cubes and outliers much. The reduction of the number of sub-cubes and outliers resulting from the splitting is greater as flattening is applied to sub-cubes down to deeper or all levels in the split procedure (large $\mu$ or $\mu = \infty$). This enhances the query performance but requires more space as the resulting sub-cubes are less dense. Flattening, when applied to only the first levels of the recursive split tree (small $\mu$), does not significantly increase the overall capacity of the resulting sub-cubes. Detailed results about the effects of flattening are given in Section 5.

## 4.2 Optional Shrinking

The continuous sub-cubes produced by the above splitting step may still not be very dense. Hence, [13] propose a step where sparse surfaces of cubes are removed. Each surface of each sub-cube is examined and its density is computed. If it is below a certain threshold $\beta$, the surface is removed and treated as a new sub-cube. This step is also repeated recursively.

In our tests for different values of $\beta$ (on a 7-dimensional real-world data cube with a capacity of $8.1 \cdot 10^8$ and roughly three million filled cells), shrinking produced a great number of additional sub-cubes (see Figure 6a), while the gain in global density was not overwhelmingly high (Figure 6b). Moreover, shrinking becomes a serious problem when dimensionality is high because more surfaces are likely to be pruned and considered as new sub-cubes. Correspondingly, the later steps will be much slower (see Figure 6c). We therefore omit the shrinking step in our approach.

## 4.3 Sub-cube Merging

In order to reduce the number of sub-cubes, nearby cubes can be merged into one if the resulting cube is not too sparse, i.e. if its density is over a certain given threshold. The result of merging a set $S$ of sub-cubes is the smallest cube encompassing all cubes in $S$, i.e. the minimum bounding region (MBR) of $S$.

The merge step is very important for our procedure, as a lower number of sub-cubes is beneficial for query performance. Hence, apparently the choice of the density threshold is one way of trading memory for query performance.

A merge step is also proposed in [3] and [13]; however, we identified some serious shortcomings there. First of all, only *pairs* of sub-cubes are considered for merging. Such a method will potentially miss candidates if, for instance, a set of $k$ sub-cubes together could be merged but no single pair of them would reach the density threshold after merging just those two. A simple example for such a situation with three sub-cubes (highlighted in gray) in two dimensions is shown in Figure 7.

If the density threshold $\chi$ for merging is chosen as, say, $\chi = 0.42$, no pair of cubes can be merged (each of the combined densities is $\leq 0.4$). However, when all three sub-cubes are considered, the resulting cube will be dense enough (0.45) and hence the cubes should be merged.

| 23  |     |    | 75  |
|-----|-----|----|-----|
| 814 |     |    | 102 |
| 89  |     |    | 60  |
|     | 512 | 4  | 31  |

**Fig. 7.** Example where a pair-wise check for merging is not sufficient

The second shortcoming is that no method is described for efficiently detecting "nearby" candidate pairs. An exhaustive comparison of all possible pairs of sub-cubes will result in a number of checks that is quadratic in the number of sub-cubes before merging. This number will of course be much higher if triples or even larger subsets of sub-cubes are also considered as candidates.

Third, the method in [13] does not check for potential overlaps of the merged cubes with other existing sub-cubes. (In fact, the merge condition given there would not even detect if an additional sub-cube were completely inside the merged area and thus increase its density – leading to the above problem.) While overlapping sub-cubes may not result in incorrect answers of queries if handled appropriately, it is surely undesirable, as reserving space for the same cells twice wastes memory. Also, some spatial index structures (such as the R+-tree) which can be used to efficiently look up relevant sub-cubes in queries, work better for non-overlapping objects.

We propose the following improved method for merging sub-cubes, which uses an important property of the R*-tree, in which the set of sub-cubes is maintained: objects in close spatial proximity are likely to be stored in the same subtree [1]. Hence, on the leaf level we can expect that nearby sub-cubes will be stored as siblings.

---

**Algorithm:** *merge(set of subcubes S ,χ)*

1.  insert sub-cubes from $S$ into R*-tree $r_1$
2.  $r_2 = $ **new** R*-tree
3.  **repeat**
4.     **for** each leaf node $l$ in $r_1$ do
5.       **if** $density(l.MBR) \geq \chi$ **and** $r_1$.intersect($l.MBR$) = $l.entries$
             **and** $r_2$.intersect(l.MBR) = $\emptyset$ **then**
6.         new sub-sube $c = $ merge($l.entries$)
7.         insert $c$ in $r_2$
8.       **else**
9.         insert $l.entries$ in $r_2$
10.      **end if**
11.    **end for**
12.    $r_1 = r_2$
13.    empty $r_2$
14. **until** (no merging during the previous *for* loop)
15. **return** set of sub-cubes in $r_1$

First, the sub-cubes resulting from the previous steps are inserted into an R*-tree $r_1$. An additional R*-tree $r_2$ is created for the newly merged sub-cubes or the sub-cubes that will not be merged. Each leaf in $r_1$ is checked for merging. For each leaf MBR (i.e. the minimum bounding hyper-rectangle containing all its entries), we compute the density (sum of sizes of the entries (sub-cubes) divided by the MBR capacity). If it is greater than or equal to the merge threshold $\chi$, we also check for over-lapping. Checking overlapping includes checking if the leaf MBR overlaps any of the newly merged sub-cubes (stored in $r_2$) or if it overlaps any entries of other leaves in its own R*-tree $r_1$. If no merging is done (the leaf MBR was not dense enough or it overlaps with other sub-cubes), the leaf entries are inserted in $r_2$ without merging. Otherwise, the leaf MBR forms a new sub-cube and is inserted in $r_2$. When all the leaves in $r_1$ have been checked (lines 4-11), we replace $r_1$ with $r_2$ and empty $r_2$. If any cubes were merged in the for-loop, the procedure is repeated. If no more merging oc-curred during the last iteration, the set of sub-cubes in $r_1$ is reported as the result of the merge. (To restrict the time taken for this step, the procedure could alternatively ter-minate after a fixed number of iterations given as an upper limit.)

This method for merging overcomes the shortcomings of the approach in [13]. In-stead of checking all pairs of sub-cubes, our method only checks candidate sub-cubes that are spatially close, based on the properties of the R*-tree. This reduces the num-ber of candidate sets to be checked in each iteration to $\lceil n/m \rceil$, where $n$ is the number of sub-cubes before merging and $m$ is the minimum number of entries in each node of the R*-tree. A brute force test of only the pairs of cubes already takes a quadratic number of checks (ignoring that the merged cubes will be compared again).

In addition, no limitation to pairs of cubes exists in our approach. If a leaf is dense enough, it will be merged regardless of how many entries it includes. So, it is left for the insertion procedure in the R*-tree to decide how many entries in a certain leaf exist.

Finally, no overlapping will occur in our merge procedure. This check is done quite efficiently through two R*-tree range queries (one each in $r_1$ and $r_2$). Otherwise, it would take linear time for each merge candidate set if it was done through checking overlaps with all the other sub-cubes.

The performance of our procedure can be affected negatively if the size of the R*-tree becomes very big and the insertion and search within the R*-tree becomes slower. This can happen if hundreds of thousands of sub-cubes are given to the proce-dure for merging. But as we have mentioned before, this problem can be handled by reducing the input set through the use of histogram flattening during the split step. Experimental results on the merging procedure are given in Section 5.

## 4.4 Filtering

Since splitting is recursive, it will often produce very small sub-cubes, many of which may still be left after merging. If the size of such a cube is below a certain threshold $\delta$, it makes more sense to consider the filled cells as outliers than to maintain a large number of small sub-cubes.

The value of $\delta$ is crucial for the balance between the number of sub-cubes and the number of outliers. The choice of this threshold depends on the strategy chosen for the representation of sub-cubes as well as on the distribution of the data. Our empiri-cal results are presented in the experiments section.

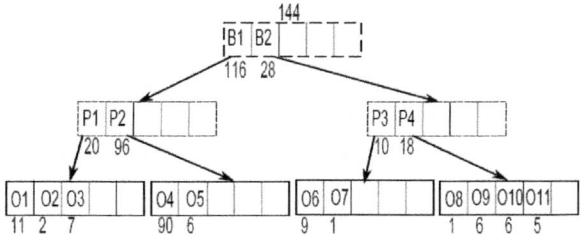

**Fig. 8.** Outliers are stored in an aggregate R*-tree (aR-tree [12])

Outliers are stored in a separate data structure. In our prototype implementation, we used an augmented version of the R*-tree also referred to as aR-tree [12]. In addition to the standard tree information, the inner nodes of the tree store aggregates of the values of their children (see. Figure 8).

## 5   Efficient Query Processing

Recall that our data structure for dense-region OLAP is a combination of two variants of the R-tree and an array-based representation of dense sub-cubes. The basic idea is to make the latter accessible through an R*-tree [1]. The outlier cells are stored in an aR-tree [12] with additional aggregate information in the inner nodes.

A range aggregate query on our proposed date structure (cf. Figure 1) will be split into two sub-queries; one within the set of sub-cubes, the other in the set of outliers (see Figure 9). In this section, we describe the how such queries can be carried out efficiently.

### 5.1   Querying Sub-cubes and Outliers

A range query in the R*-tree returns a list $L$ of all the sub-cubes that intersect with the query range. The query is then translated into the appropriate range query for each cube in $L$. Our representation of these cubes will answer each of these queries efficiently. The combined aggregate of the results is the answer of the first sub-query.

The second sub-query is carried out on the outlier cells, which are stored in the leaf nodes of the aR-tree (see Figure 8). A range query in this augmented tree does not have to return all cells in the query range. Instead, if the MBR of a whole subtree is contained in the query range, the aggregate value stored in the inner node pointing to this subtree can be used, making the query more efficient. In the example in Figure 8, if the MBR of entry P4 is fully contained in the range of a query, the aggregate value 18 stored with P4 is used directly instead of descending further down the tree.

### 5.2   Parallel Computation

A very basic and straightforward way of further enhancing query processing is the parallel computation of the two parts of a query in separate threads, as depicted in Figure 10.

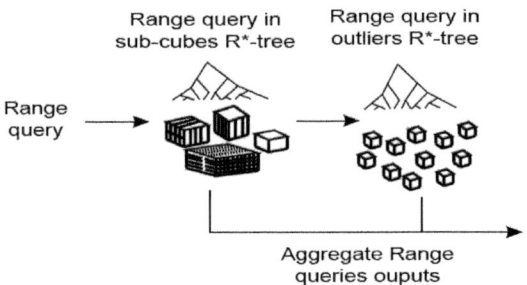

**Fig. 9.** Processing a range query on the hybrid data structure

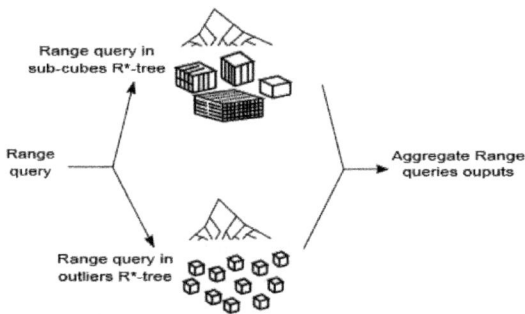

**Fig. 10.** Parallel execution of both sub-queries can enhance query performance

Of course the workload in an arbitrary query will often not be divided evenly between the two sub-queries so one thread will have to wait for the other to finish before the actual result of the query can be obtained. However, since queries usually arrive in bulks (because users request views consisting of hundreds or thousands of range queries), each thread can process its share of all queries before the final result is created. We have achieved significant speedups in some of our tests using this strategy, which are described in the next section.

## 6   Experimental Results

In this section, we report our experiments carried out to test the methods described in the previous sections. Many experiments reported in other literature such as [13] used artificial and usually small data sets. In our experiments, real data cubes provided by an industrial partner specializing on business intelligence were used. The data cubes were anonymized for privacy but with no change to the model and the filled positions of the data cube. The data cubes used in these experiments vary in all or some of properties like dimensionality, size and capacity. Table 1 lists the properties of these data cubes.

Except $DC_1$, a dense cube used for comparison, all cubes are rather sparse; $DC_2$ and $DC_5$ can be considered very sparse. All of the cubes have dimensionalities that are common in practical OLAP scenarios.

**Table 1.** The data cubes used in the experiments

| Name | D | Size | Capacity | Density |
|---|---|---|---|---|
| $DC_1$ | 6 | 136,558 | 171,000 | $\approx 0.7986$ |
| $DC_2$ | 9 | 184,775 | $\approx 3.9 \cdot 10^{14}$ | $\approx 4.7 \cdot 10^{-10}$ |
| $DC_3$ | 9 | 8,627,324 | $\approx 5.4 \cdot 10^{9}$ | 0.0016 |
| $DC_4$ | 7 | 3,069,183 | $\approx 8.1 \cdot 10^{8}$ | 0.003 |
| $DC_5$ | 13 | 3,154,857 | $\approx 6.1 \cdot 10^{21}$ | $\approx 5.2 \cdot 10^{-16}$ |

For our tests of query performance, for each data cube a set $Q_{1d}$ of 10,000 queries was generated such that each query range was a slice of the cube along one dimension. The dimension where the slicing occurred and the slice width were chosen randomly. The queries were intentionally chosen as "bad cases" with big query ranges, as this requires more extensive lookups in the R*-trees and each query will involve more than just one or a few sub-cubes. Still, this kind of slicing will not always produce equal-sized ranges, since the sizes of the dimensions are different and each dimension is equally likely to be sliced.

Although these queries are not from real-world scenarios (actual queries were not available to the authors), each one of them reflects the structure of an aggregated value queried in an OLAP view in a scenario with predefined hierarchies. We assume that – whenever possible – hierarchies are represented such that base elements belonging to the same subtree of a hierarchy tree will be located in a contiguous range in the internal SDDC representation (cf. Figures 3 and 4). If there are multiple independent hierarchies in one dimension, this may of course not always be possible, and query performance might degrade with respect to some hierarchical elements, as the query will have to be split up into several partial queries.

Note that since our focus is on the different parameters of the refined sub-cube extraction, all experiments regarding query performance consist of relative comparisons conducted to explore performance improvements due to parameter settings in our dense region extraction. The output of the extraction step – i.e. the set of dense regions and the set of outlier cells – could be used as input to other dense-region-based OLAP approaches to test performance changes there as well. In order to see absolute performance comparisons with other systems, further tests need to be done.

In each of the following experiments, one of the above steps or methods will be illustrated. An Intel DualCore 2 CPU T7600 with 2 GB RAM was used to run the experiments. In all the experiments of this chapter the minimum histogram value $\alpha$ was given the value 0.01; the reason is that when flattening is applied (as was the case in most of our tests), we wanted to consider any non-zero histogram values as part of the dense intervals. Unless stated otherwise, for all tests the flattening level threshold $\mu = 1$ was used, i.e. flattening was only used during the initial split of the original data cube. Rather than choosing one fixed flattening factor $f$, this parameter was adapted to the respective dimension length and set as $f(D_i) = \log_{10} n_i$, a value which proved useful in preliminary tests.

Cube $DC_1$ was tested only in the experiments reported in the next section. The reason is that it only produces two sub-cubes after splitting, and these two later merge into a single sub-cube. Hence, there is no need for flattening and there are no outliers, which means that no speedup will occur in the parallel computation of queries.

The pre-aggregation strategy used in the IDC representation of the dense cubes was uniformly chosen as *prefix sum*, such that each range query within a sub-cube is carried out in constant time (cf. [10]).

## 6.1   Overall Effects of the Dense Sub-cube Extraction

In this subsection, our goal is to show how the extraction procedure can reduce the sparsity to a great extent, which makes it possible to keep data cubes (even large ones) in main memory. In this experiment, all data cubes were used (except $DC_5$, which is described in more details later).

The thresholds used for all the data cubes in this section are the same. For assessing the new overall capacity, we only looked at the sum of capacities of the sub-cubes (the linear space overhead of the R*-tree is ignored) here. Also, in this step no filtering was done in order to include all the filled cells in the sub-cubes and not to have any outliers. A merge threshold $\chi = 0.5$ was used, so that merging would only be done if the newly merged sub-cubes had densities greater or equal to 0.5. For $DC_1$, no merging was applied because only two sub-cubes will be given to the merge procedure.

Table 2 shows the space required by the full data cubes before and after extracting the sub-cubes:

**Table 2.** Capacities before/after the extraction procedure

|  | $DC_1$ | $DC_2$ | $DC_3$ | $DC_4$ |
|---|---|---|---|---|
| **Original data cube capacity** | 171,000 | $\approx 3.9 \cdot 10^{14}$ | $\approx 5.4 \cdot 10^9$ | $\approx 8.1 \cdot 10^8$ |
| **Sum of capacities of sub-cubes** | 164,160 | 361,033 | 25,078,968 | 5,257,635 |

The changes in (global) density through the sub-cube extraction can be seen in Figure 11. Densities are expressed as percentages.

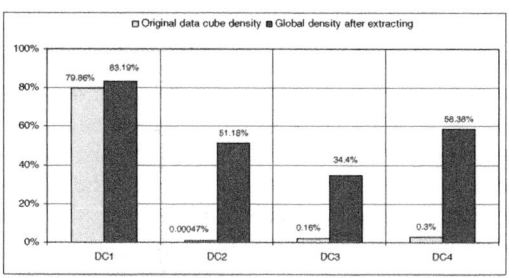

**Fig. 11.** Global densities before and after extraction

Obviously, the density is increased dramatically, even though the merge step often introduces some sparsity again.

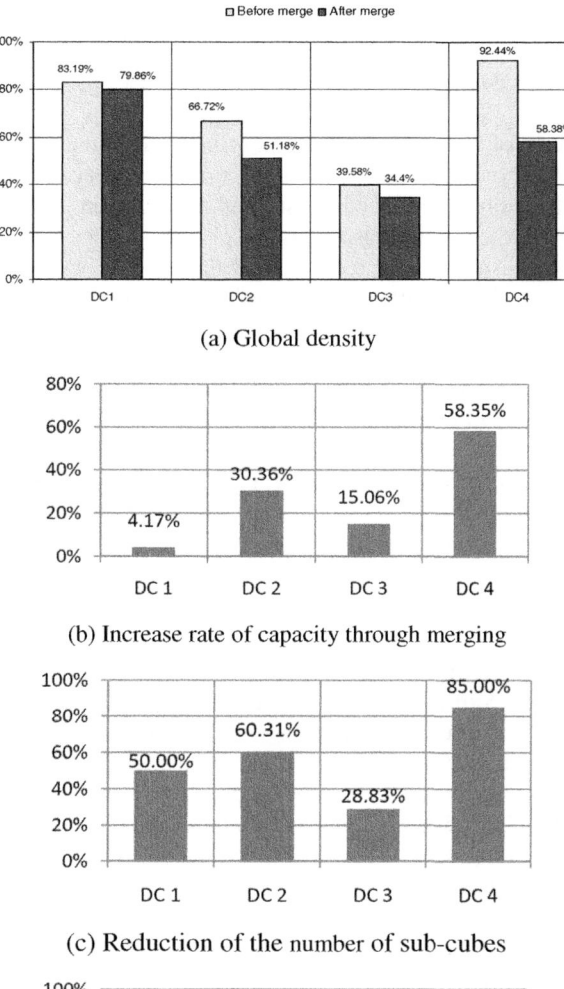

(a) Global density

(b) Increase rate of capacity through merging

(c) Reduction of the number of sub-cubes

(d) Query speedup through merging

**Fig. 12.** Effects of merging sub-cubes after splitting

## 6.2  Effects of Merging

In this part, we compare the results of the extraction procedure before and after the merge step. Since we only aimed to see the effects of merging, no filtering was done to these data cubes. The threshold used for merging is $\chi = 0.5$ (the newly merged sub-cubes should be at least 50% dense). $DC_5$ has been ignored in this experiment and will be explained in more detail in the section 5.4.4. Figure 12 provides a synopsis of the effect of merging.

Let $S_b$ be the set of sub-cubes before merging and $S_a$ the set of sub-cubes after merging. Figure 12 shows the results of our experiments. As can be seen in Figure 9a, merging causes the global density of the sub-cubes to decrease. This is a rather obvious result that was expected because merging closely located sub-cubes will usually add sparse regions. However, this reduction of density is not nearly as drastic as the gain in density due to the overall extraction (cf. Figure 11), which is mainly due to the splitting step. The corresponding increase in overall capacity is shown in Figure 12b.

Apparently, merging can significantly reduce the number of sub-cubes (which is the main goal of this step). As can be seen in Figure 12c, a reduction between 28% and 85% was achieved in our tests. (This number can of course be increased by lowering the merge threshold, if memory is available.)

The effects of this reduction on query performance are significant, as shown in Figure 9d. The speedup in queries is almost proportional to the reduction rate. The query speedup $s$ is measured as follows:

$$s = 1 - \frac{t(S_a, Q)}{t(S_b, Q)}$$

where $t(S, Q)$ is the time (in seconds) needed to do the set $Q$ of queries on a data cube represented by the set $S$ of sub-cubes. The set used as $Q$ was the generated query set $Q_{1d}$. It is interesting to see how the reduction of time for the queries corresponds closely to the reduction in the number of sub-cubes. So, significant speedup can be obtained through a reduction of sub-cubes, achieved in a tradeoff with extra space. One might expect $DC_1$ to give a better speedup ratio than $DC_3$, because it had a bigger sub-cubes reduction rate. The reason it did not occur, is that in $DC_1$ the number of sub-cubes was just reduced from two to one through merging, which hardly changed the sub-cubes R*-tree size. Also, the time needed for the querying was already very short because of this (around 0.4 seconds for 10,000 queries), which makes it harder to achieve a high speedup.

## 6.3  Histogram Flattening

To show all effects of flattening mentioned above, we performed a set of experiments on $DC_5$. In these experiments, we applied flattening using different values of the flattening level threshold $\mu$. The horizontal axis in all figures of this section represents different values for $\mu$. Again, $\mu = 0$ means no flattening; $\mu = 1$ means that flattening is applied only to the original cube; and so on. This data cube ($DC_5$) produces a lot more sub-cubes than the other data cubes (see Figure 13a, where $\mu = 0$). Hence, in order to reduce this large number of sub-cubes, we aimed to increase the number of merges by chose a very low merge threshold $\chi = 0.01$. In this section, we will see how flattening can also help solve this problem.

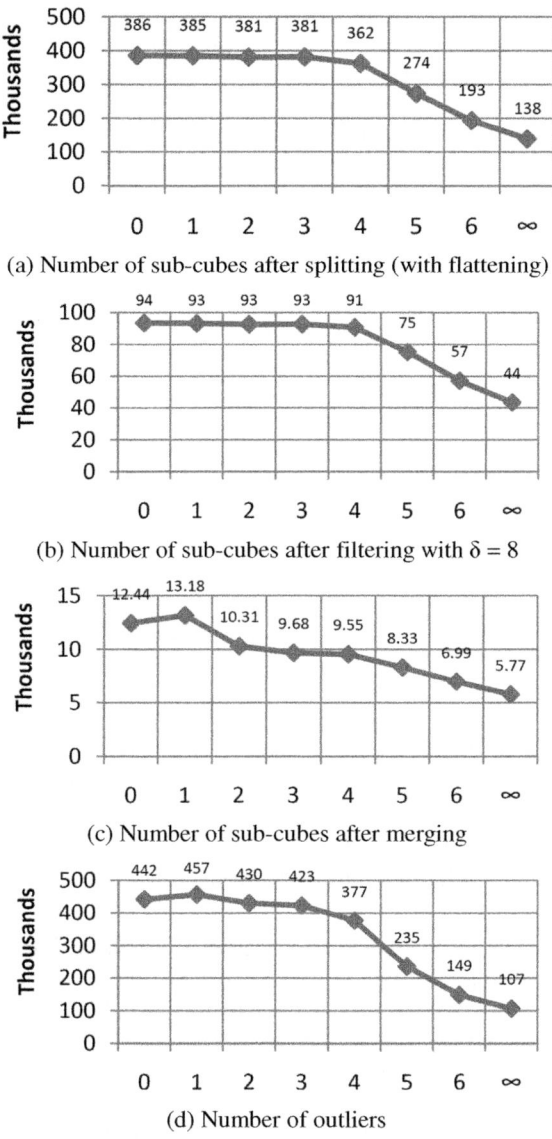

(a) Number of sub-cubes after splitting (with flattening)

(b) Number of sub-cubes after filtering with δ = 8

(c) Number of sub-cubes after merging

(d) Number of outliers

**Fig. 13.** Effects of flattening with different values for $\mu$

Figure 13 illustrates the change in the number of sub-cubes after each step and the final number of outliers after merging. As shown in this figure, the number of sub-cubes and outliers decreases as the flattening is applied down to deeper levels in the split procedure. From Figures 13a and 13b, we can see that the reduction rate in the number of sub-cubes between higher values for $\mu$ (4, 5, 6 and ∞) is larger compared to the reduction rate between the lower values 0, 1, 2 and 3. The reason why the number of outliers goes down with deeper flattening is that many outliers are "swallowed" by the larger sub-cubes which are produced.

As a side node, Figure 13c shows that the merge procedure does not provide a guaranteed behavior, as we can see with $\mu$ values 0 and 1. When $\mu = 1$, one would normally expect fewer sub-cubes after merging than for $\mu = 0$ because it took fewer sub-cubes as input. But this not necessarily the case (as in our example) because there is no guarantee of how the sub-cubes will be inserted in the R*-tree. The same applies to Figure 13d at the same positions, where the number of outliers is expected to be smaller, as fewer sub-cubes are expected (due to merging).

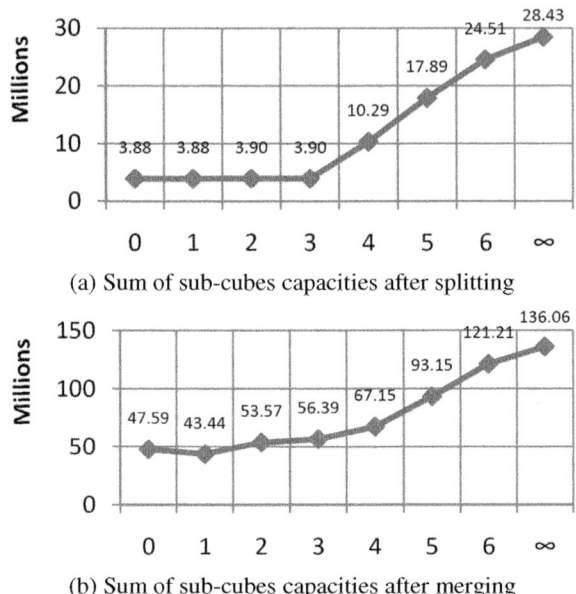

(a) Sum of sub-cubes capacities after splitting

(b) Sum of sub-cubes capacities after merging

**Fig. 14.** Effects of flattening on capacities

Figure 14 shows how the capacities increase as the flattening is applied further to deeper levels of splitting. The rate at which the capacities increase, becomes higher as the value of $\mu$ increases. However, the capacity is still far away from that of the original data cube ($\approx 6.1 \cdot 10^{21}$).

The observation that flattening will only increase the capacity of the sub-cubes significantly when it is applied down to deeper levels, can be explained as follows: the increase of capacity is expected to mostly occur from the sub-cubes returned by the split procedure at splitting levels where flattening was applied. Flattening will increases the capacity a sub-cube because our minimum histogram threshold $\alpha = 0.01$ considers any non-zero histogram cell as part of a dense interval. So, if flattening was applied with $\mu = 1$, the sub-cubes that are expected to increase the sum of the capacities are the sub-cubes reported (as continuous) from the first level ($l = 1$)in the recursive split tree. However, reporting continuous sub-cubes in the splitting procedure is most likely to occur at deeper levels in the recursive split tree. This is why applying flattening down to deeper levels as opposed to only the first levels is more likely to affect the capacities of the sub-cubes.

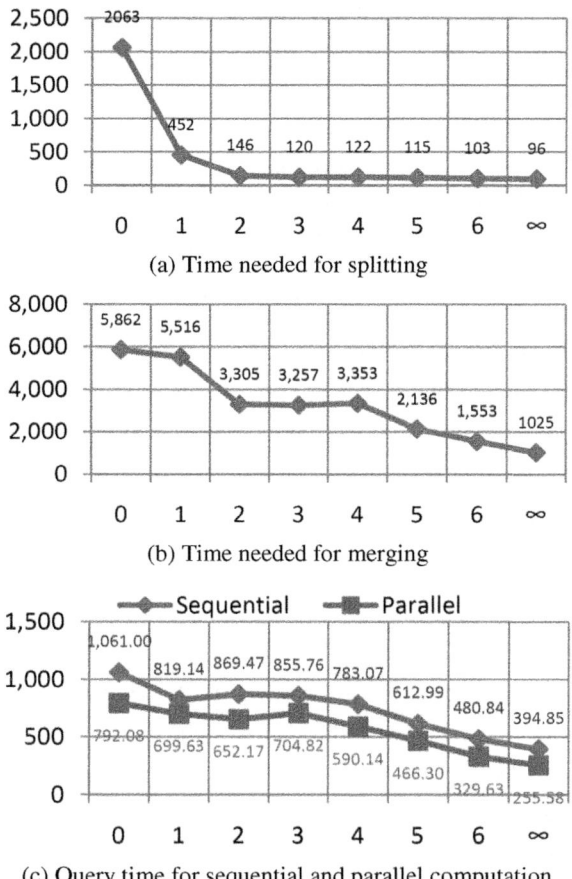

(a) Time needed for splitting

(b) Time needed for merging

(c) Query time for sequential and parallel computation

**Fig. 15.** Effects of flattening on performance

Figure 15 shows how flattening can help in enhancing the time efficiency of the extraction procedure (Figure 15a and 15b) and for the query performance (Figure 15c). The most significant effect is on the speed of the splitting procedure, where flattening restricted to the top level ($\mu = 1$) is sufficient for a reduction of the time by more than 75%. The reason is that the first split (without flattening) produces a lot of sub-cubes and the splitting procedure has to check for each newly created sub-cube whether it already exists. Hence, the effect of applying the flattening at the first level gives a dramatic speedup in the splitting step. Still, as can be seen in Figure 13, there is no significant change in the quality of the output.

Flattening has a strong effect on the efficiency and effectiveness of our extraction procedure. If it is applied with a small value for $\mu$, we can expect a faster splitting with no significant change in the output of the procedure. Hence, we advice to use flattening at least with flattening level $\mu = 1$. If memory limits are less important, flattening gives us the ability to use some extra space to reduce the number of sub-cubes after splitting; this speeds up the extraction procedure (especially the merge step) and most importantly enhances the performance of queries. Hence, the flattening level threshold $\mu$ is another way to adjust the tradeoff between query time and space.

## 6.4 Sub-cubes vs. Outliers

In another set of experiments, we investigated different tradeoffs between the number of outliers and the number of sub-cubes and check the performance using each of these different combinations. The data cubes tested here are $DC_2$, $DC_3$, $DC_4$ and $DC_5$. In data cubes $DC_2$, $DC_3$ and $DC_4$, the merge threshold used was $\chi = 0.5$. A different merge threshold was used for $DC_5$, since the number of sub-cubes resulting from splitting was very large (384,966 sub-cubes). So, for $DC_5$, a different merge threshold was used, and filtering was applied before the merge step to reduce the number of sub-cubes given to the merge procedure.

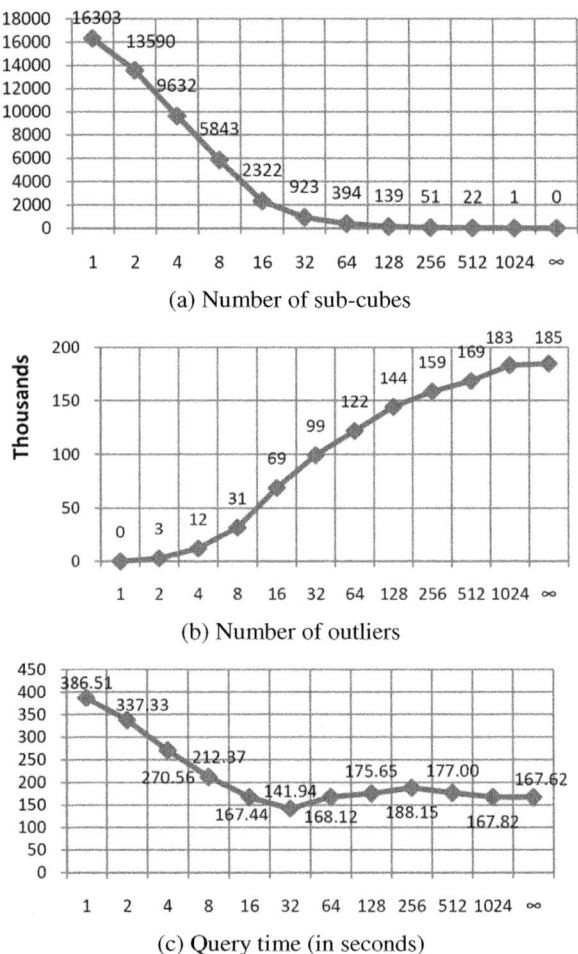

(a) Number of sub-cubes

(b) Number of outliers

(c) Query time (in seconds)

**Fig. 16.** Effects of filtering in data cube $DC_2$

The balance between the number of sub-cubes and the number of outliers was changed through different filtering thresholds. So, different values of the *minimum size* $\delta$ were used. The horizontal dimension in all the graphs in this section corresponds to different $\delta$ values. The larger the value of $\delta$, the more sub-cubes will be discarded and the more outliers will be produced. $\delta = 1$ means that no filtering is done, whereas $\delta = \infty$ turns all filled cells into outliers.

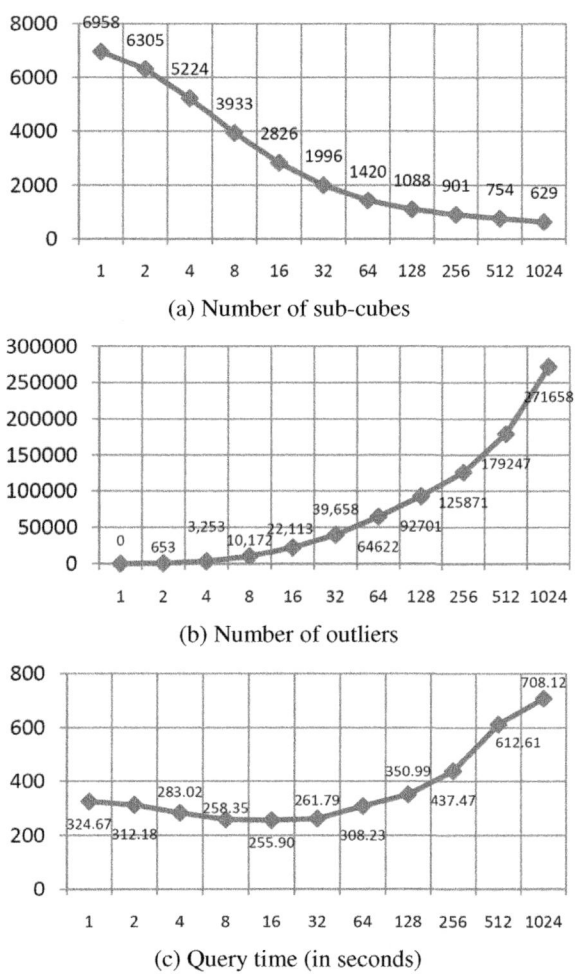

(a) Number of sub-cubes

(b) Number of outliers

(c) Query time (in seconds)

**Fig. 17.** Effects of filtering for cube $DC_3$

### 6.4.1 Results for $DC_2$

Data cube $DC_2$ resulted in a big number of sub-cubes considering its size. One can expect then that a lot of these sub-cubes are very small and will therefore be discarded during filtering. The result shown in Figure 16 confirms this assumption. In parts (a) and (b), we can see how filtering using small values for $\delta$ (1, 2, 4, 8 and 16) discards a large number of sub-cubes and classifies their filled cells as outliers. In $DC_2$, the best query performance was achieved when $\delta = 32$. This confirms what was said before; small sub-cubes should be discarded and considered as outliers because no speedup occurs from explicitly storing such small sub-cubes. This is why in this data cube, which contains mostly small sub-cubes, considering all filled cells as outliers is also one of the possible configurations that results in a good query performance.

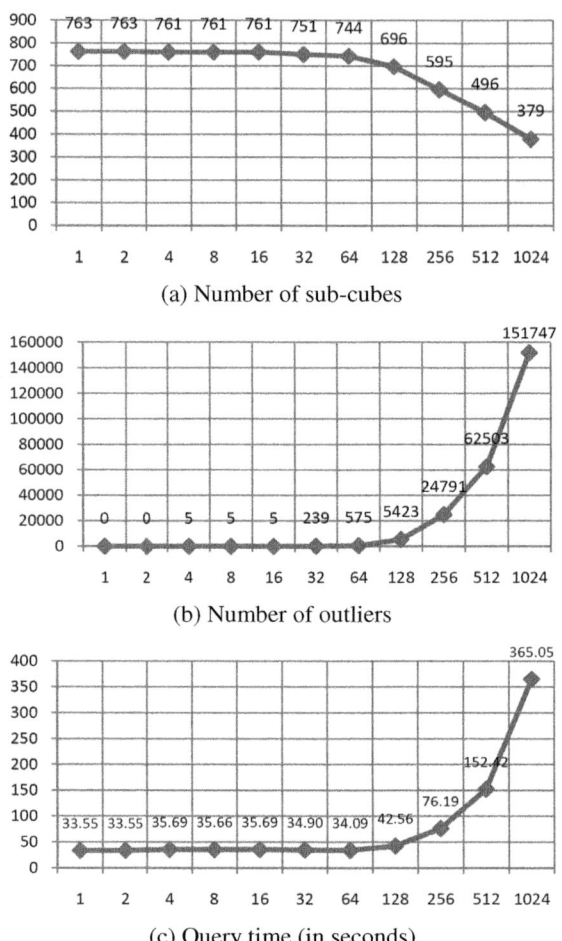

(a) Number of sub-cubes

(b) Number of outliers

(c) Query time (in seconds)

**Fig. 18.** Effects of filtering for $DC_4$

### 6.4.2  Results for $DC_3$

For $DC_3$, the same idea was tested, but the case $\delta = \infty$ was not tested due to the size of the data cube (over 3 million filled cells), the same holds for $DC_4$ and $DC_5$.

As can be seen in Figure 17, $\delta$ is a good tuning mechanism to control the number of sub-cubes and outliers. The cube contains larger sub-cubes (629 with size greater than 1024) and is a good example of how a good balance between outliers and sub-cubes enhances time efficiency. In this case, at $\delta = 16$ the ratio between number of sub-cubes and number of outliers yielded the best query performance (cf. Figure 17c).

### 6.4.3 Results for DC$_4$

DC$_4$ is different from the other tested data cubes, as the number of sub-cubes after splitting is not too high (compared to its size). This means that probably not too many of these sub-cubes will be small.

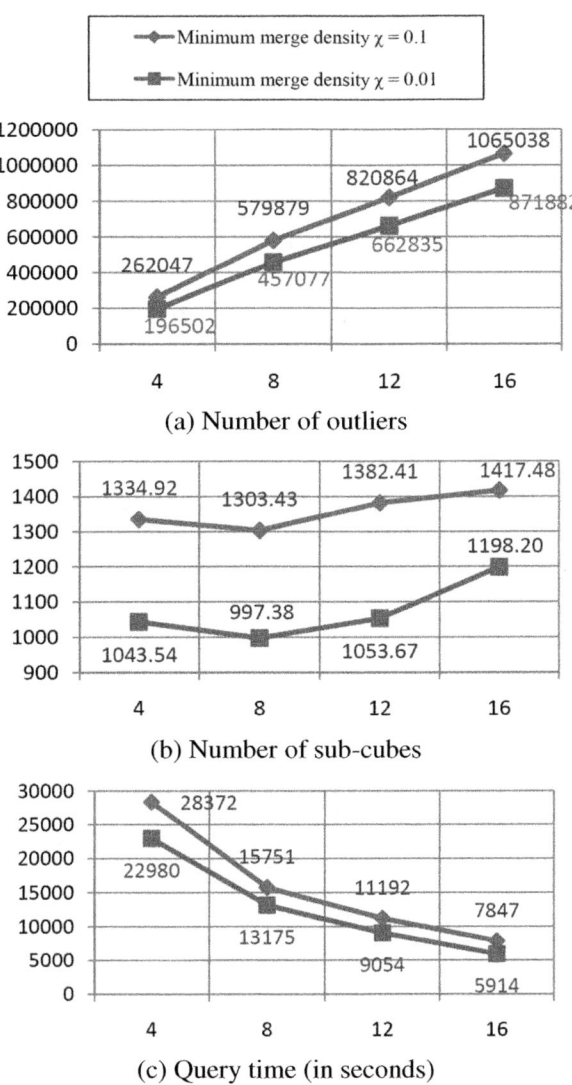

(a) Number of outliers

(b) Number of sub-cubes

(c) Query time (in seconds)

**Fig. 19.** Effects of filtering for DC$_5$

As illustrated in Figure 18a and 18b, hardly any changes occurred when filtering with small values for $\delta$. This means most sub-cubes are relatively big. This is why no or few outliers give the best performance, but if the number of outliers becomes too large (some big sub-cubes are discarded), the query performance will drop, as can be seen in Figure 18c).

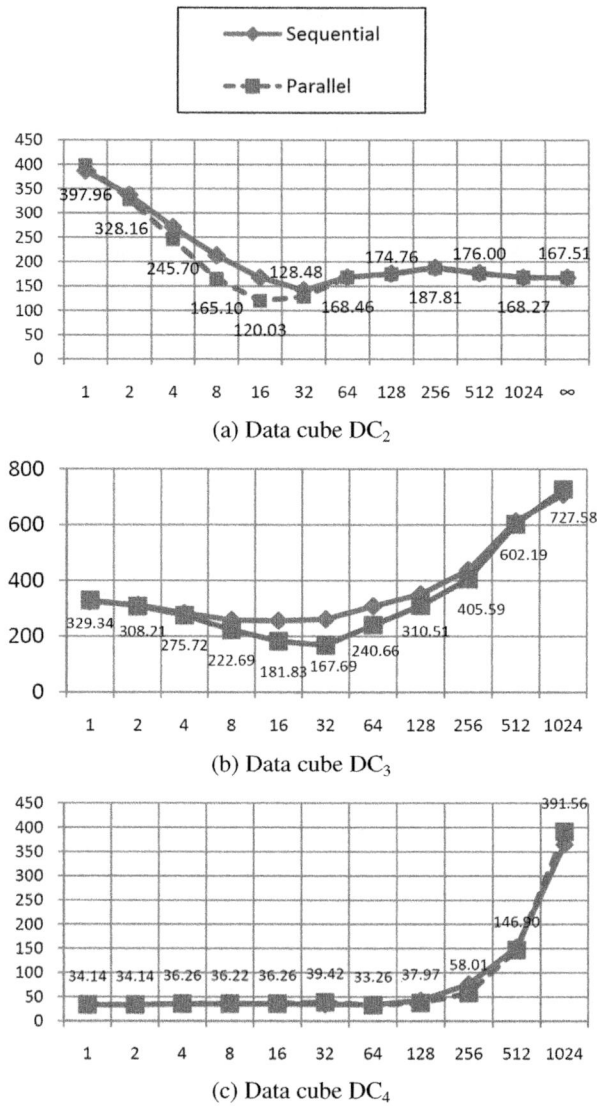

**Fig. 20.** Comparison between sequential and parallel computation of queries in different data cubes

### 6.4.4  Experimental Results for DC$_5$

As mentioned before, DC$_5$ resulted in a lot of sub-cubes from the splitting step (384,966 sub-cubes), indicating that the sparsity is distributed rather evenly in this cube. To reduce the number of sub-cubes given as an input to the merge step, filtering was performed before rather than after the merging. Also, two very low merge thresholds were used in the tests ($\chi = 0.1$ and $\chi = 0.01$), which helps reduce the number both of sub-cubes and outliers at the same time, at the cost of increased space requirements. Our tests results for this data cube are shown in Figure 19.

We can see the effect of merging mentioned before (fewer sub-cubes, fewer outliers and better query time). In this example, the best query time was achieved when $\delta = 8$. This is where we obtained a good balance between sub-cubes and outliers. Still, as we showed before, if memory is not an issue, flattening can be very useful and can be applied to bigger $\mu$ values (not only to the first level, as was done here).

One important result from all these tests in this section is that when the number of the resulting sub-cubes is very large relatively to the size of the data cube (i.e. it will most likely include a lot of small sub-cubes), it is advisable to apply filtering with suitable thresholds to discard these small sub-cubes, in spite of the fact that this will increase the number of outliers. To have a good performance, the sub-cubes should be relatively big and a good balance between the number of sub-cubes and outliers should be found. This includes the usual case where larger sub-cubes and a number of outliers exist (for example DC$_3$) and two other possible cases where there are either only relatively big sub-cubes and few or no outliers (for example DC$_4$), or almost only outliers and no big sub-cubes (for example DC$_2$).

### 6.5  Parallel Computation

Several experiments were performed to test the effects of carrying out sets of queries in parallel. First, the set $Q_{1d}$ of queries was executed sequentially, then a parallel version of the computation was carried out in two separate threads, where one thread computed the outputs from the sub-cube R*-tree and the other thread from the outlier R*-tree. The cubes DC$_2$, DC$_3$, DC$_4$ and DC$_5$ were tested. Again, the horizontal axis represents the different values of filtering threshold *minimum size* $\delta$. The results are shown in Figure 20.

As can be seen in Figure 20, if either (almost) only sub-cubes or only outliers exist, the parallel version will of course not provide any speedup. This also explains why no speedup occurs for $\delta = 1$ (no filtering, just sub-cubes) and $\delta = \infty$ (all filled cells are outliers). Also, if filtering is done with small $\delta$ values and not a lot of outliers are produced, no speedup is likely to occur. Some data cubes did not give any significant speedup regardless of the number of outliers, like in DC$_4$. The reason is that by having only (large) sub-cubes, we gained the best performance. However, when filtering enhanced the time in the sequential version (like in DC$_2$ or DC$_3$), the parallel version always seemed to give a better speedup.

An important observation is that even if the parallel version does not necessarily enhance the performance at the optimal point of the sequential computation, the parallel version is never slower at that point. Thus, if we have a good configuration in the sequential version of the program with both outliers and sub-cubes, the parallel version is expected to give a better or at least the same performance (but not worse).

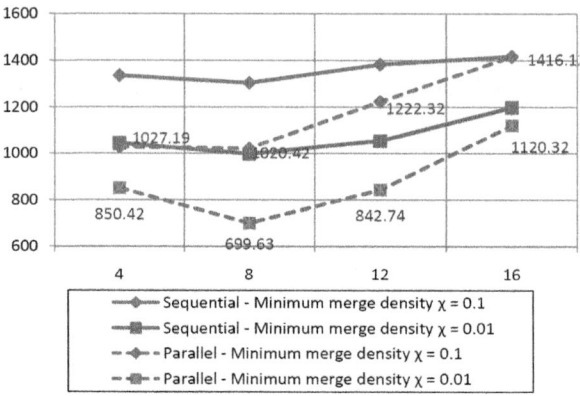

**Fig. 21.** Sequential vs. parallel queries in DC$_5$

For data cube DC$_5$, the sparsest one in our test, the parallel computation of queries showed the biggest effects, as can be seen in Figure 21. For example, when the minimum size $\delta = 8$ and the minimum merge density $\chi = 0.01$, the time required by the parallel version is about 30% less than for the sequential version. Hence, our approach seems to be effective especially in cases with high sparsity of the original cube and a low merge threshold (where space is traded in for query performance).

# 7 Conclusions

We have presented a dense-region-based data structure for in-memory OLAP, which can be constructed a lot more efficiently than earlier approaches. It uses an efficient data structure for representing dense sub-cubes, which are indexed through an R*-tree. An augmented version of this, the aR-tree, is used to maintain and efficiently query outlier cells, which do not belong to any dense region.

The resulting global density of the new representation can be influenced by two parameters: the level of histogram flattening and the minimum merge threshold. This allows for an adjustable trade-off between query performance and memory usage. In addition, a filtering threshold can be used to fine-tune the ratio of sub-cubes to outliers for faster query processing. Tests with several real-life data cubes have confirmed the usefulness of our approach, especially for very sparse cubes, and have shown good sample choices of the parameters for different cubes.

The most important work remaining involves a possible automation of the selection of parameter values on the basis of heuristics derived from our tests, which will greatly improve the usability of the approach when integrated in commercial OLAP systems. Also, while our data structure for representing dense sub-cubes is very efficient for cell updates (which are required when OLAP cubes are used in planning or "what-if" scenarios), in our experiments we did not take into account updates of cube cells.

Finally, parallelizing the computation has only been touched upon in a very straightforward manner for query processing. It could also be utilized in other parts of our approach, e.g. during sub-cube extraction. We are currently working on a different approach using GPUs for massively parallel OLAP computation.

**Acknowledgments.** The research presented in this article was supported by the German Research Foundation (DFG) within its knowledge and technology transfer program and by Jedox AG. The authors would like to thank Philippe Hagedorn, Zurab Khadikov, Dominic Mai, and three anonymous reviewers for valuable comments and criticism.

# References

1. Beckmann, N., Kriegel, H.-P., Schneider, R., Seeger, B.: The R*-tree: An efficient and robust access method for points and rectangles. In: Proceedings of ACM SIGMOD, pp. 322–331 (1990)
2. Cheung, D.W., Zhou, B., Kao, B., Kan, H., Lee, S.D.: Towards the building of dense-region-based OLAP system. Data and Knowledge Engineering 36(1), 1–27 (2001)
3. Chun, S., Chung, C.-W., Lee, S.-L.: Space-efficient cubes for OLAP range-sum queries. Decision Support Systems 37(1), 83–102 (2004)
4. Cuzzocrea, A., Wang, W.: Approximate range-sum query answering on data cubes with probabilistic guarantees. Journal of Intelligent Information Systems 28, 161–197 (2007)
5. Geffner, S., Agrawal, D., El Abbadi, A., Smith, T.: Relative prefix sums: an efficient approach for querying dynamic OLAP data cubes. In: Proceedings of International Conference on Data Engineering, Sydney, Australia, pp. 328–335 (1999)
6. Gray, J., Chaudhuri, S., Bosworth, A., Layman, A., Reichart, D., Venkatrao, M., Pellow, F., Pirahesh, H.: Data cube: A relational aggregation operator generalizing group-by, cross-tab, and sub-totals. Data Mining and Knowledge Discovery, 29–53 (1997)
7. Gupta, H., Harinarayan, V., Rajaraman, A., Ullman, J.: Index selection for OLAP. In: Proceedings of the 13th International Conference on Data Engineering, pp. 208–219 (1997)
8. Guttman, A.: R-trees: A dynamic index structure for spatial searching. In: Proceedings of ACM SIGMOD, pp. 47–57 (1984)
9. Haddadin, K., Lauer, T.: Efficient Online Aggregates in Dense-Region-Based Data Cube Representations. In: Proceedings of DaWaK, Linz, Austria, pp. 177–188 (2009)
10. Ho, C.-T., Agrawal, R., Megido, N., Srikant, R.: Range queries in OLAP data cubes. In: Proceedings of ACM SIGMOD, pp. 73–88 (1997)
11. Lauer, T., Mai, D., Hagedorn, P.: Efficient range-sum queries along dimensional hierarchies in data cubes. In: Proceedings of the First International Conference on Advances in Database, Knowledge, and Data Applications, Cancún, Mexico, pp. 7–12 (2009)
12. Mamoulis, N., Bakiras, S., Kalnis, P.: Evaluation of top-k OLAP queries using aggregate R-trees. In: Bauzer Medeiros, C., Egenhofer, M.J., Bertino, E. (eds.) SSTD 2005. LNCS, vol. 3633, pp. 236–253. Springer, Heidelberg (2005)
13. Lee, S.-L.: An effective algorithm to extract dense sub-cubes from a large sparse cube. In: Tjoa, A.M., Trujillo, J. (eds.) DaWaK 2006. LNCS, vol. 4081, pp. 155–164. Springer, Heidelberg (2006)
14. Riedewald, M., Agrawal, D., El Abbadi, A.: Flexible data cubes for online aggregation. In: Proceedings of the 8th International Conference on Database Theory, London, UK, pp. 159–173 (2001)

15. Riedewald, M., Agrawal, D., El Abbadi, A.: pCube: Update-efficient online aggregation with progressive feedback and error bounds. In: Proceedings of the 12th International Conference on Scientific and Statistical Database Management, Berlin, Germany, pp. 95–108 (2000)
16. Witten, I.H.: Data Mining: Practical Machine Learning Tools and Techniques. Addison-Wesley, Reading (2000)
17. Zhao, Y., Deshpande, P., Naughton, J.: An array-based algorithm for simultaneous multidimensional aggregates. In: Proceedings of ACM SIGMOD, Tucson, AZ, pp. 159–170 (1997)

# Improving Access to Large Patent Corpora

Richard Bache and Leif Azzopardi

University of Glasgow, Department of Computing Science,
Glasgow G12 8QQ, UK
{bache,leif}@dcs.gla.ac.uk

**Abstract.** Retrievability is a measure of access that quantifies how easily documents can be found using a retrieval system. Such a measure is of particular interest within the patent domain, because if a retrieval system makes some patents hard to find, then patent searchers will have a difficult time retrieving these patents. This may mean that a patent searcher could miss important and relevant patents because of the retrieval system. In this paper, we describe measures of retrievability and how they can be applied to measure the overall access to a collection given a retrieval system. We then identify three features of best-match retrieval models that are hypothesized to lead to an improvement in access to all documents in the collection: sensitivity to term frequency, length normalization and convexity. Since patent searchers tend to favor Boolean models over best-match models, hybrid retrieval models are proposed that incorporate these features while preserving the desirable aspects of the traditional Boolean model. An empirical study conducted on four large patent corpora demonstrates that these hybrid models provide better access to the corpus of patents than the traditional Boolean model.

## 1 Introduction

Patent searches are recall-dominated as there is a significant cost associated with failing to retrieve relevant documents [15]. A recently proposed measure, retrievability [6–8, 11], is an attribute of particular interest in the patent domain. Essentially, the retrievability of a document is the ease with which that document can be retrieved – and the retrievability of a document depends upon the document collection and the Information Retrieval (IR) system used. A document with low retrievability is likely to be very difficult, if not impossible to find, while a document with high retrievability is likely to be much easier to find. For a given corpus, different IR systems will yield different levels of retrievability across the population of documents. It is therefore important to select an IR system that ensures that all documents are as accessible as possible. This is particularly the case in patent retrieval; because if patent searchers employ IR systems that limit their ability to retrieve particular documents in the collection, this could mean missing relevant documents. Early work using retrievability has provided interesting insights into the problem of documents accessibility [6]. In [8] it was shown that different best match retrieval systems provided substantially different levels

A. Hameurlain et al. (Eds.): TLDKS II, LNCS 6380, pp. 103–121, 2010.

of retrievability across a document collection, while in [10, 11] it was shown that pseudo-relevance feedback can skew the retrievability of documents (i.e. some documents become much more retrievable than others). With such variations in access to documents arising due to the retrieval system, it is important to quantify and understand its influence on the retrieval of documents.

In this work, we consider the influence of different retrieval systems on large patent corpora, and determine the retrievability of patents when using such systems. The overall access to the patent corpora is determined to provide an indication of the how accessible the population of documents is for each given system. Since patent search is often conducted using traditional Boolean systems (i.e. exact match retrieval models), we shall examine these types of retrieval models, and compare them against best-match retrieval models. Best-match models have been favored in the IR research community because they have been shown to deliver excellent retrieval performance. However within the patent domain, searchers prefer exact-match models because of custom and tradition, the precise interpretation of boolean queries, and the legal and regulatory requirements that are often imposed. To improve the access of exact-match models, while preserving these required features, we also consider a series of hybrid retrieval models. These models accept a Boolean query and provide a crisp cut-off between retrieved and non-retrieved documents, as the traditional Boolean model already does, but incorporate a number of features of best-match models that improve the access to the collection. An empirical study using the MAREC patent test collection builds on a previous pilot study [9] and provides provides evidence that these features often improve access (i.e. they make access to individual documents more equal). It is shown that when these features are incorporated within the hybrid models this leads to improved access over the traditional Boolean model.

The rest of the paper is organized as follows. In Section 2, we summarize the reasons why those in the patent domain prefer Boolean queries and, by implication, models which accept queries in this form. Section 3 formally defines the concept of retrievability and describe how it can be measured. Such measurement requires a large number of representative queries so in Section 4, we consider both the quality and quantity of queries. Section 5 identifies and explains the three features that were hypothesized to increase access. In Section 6 we give formal definitions of the variants of Tf-Idf and BM25 used for this study and then define the hybrid models. Then, Section 7 presents the results of the empirical study that we conducted to analyze these models on a number of different patent corpora. This is followed by Section 8 which concludes with a summary of findings and directions for future work.

## 2    Patent Searching

A patent searcher typically requires accessing large corpora, consisting of millions of documents, in order to perform a variety of search tasks [15]. Some common search tasks include:

**Novelty Search:** given a patent application, the search task is to ensure that the claims made of the new invention have not been previously patented or documented.

**Validity/Invalidity Search:** The search task is to investigate existing patents to determine whether their claims are enforceable, or to determine whether any other patents violate an existing or currently held patent (validity/ invalidity search).

**Freedom to Operate:** A search is instigated to determine if a proposed course of action violates an existing patent.

To accomplish such tasks, patent searchers often prefer exact-match models where the query is submitted in a Boolean form using the AND, OR and NOT operators [5, 12]. In response to such a query, a system employing an exact-match model, will return all the documents for which the query is true. Since the documents returned are not ranked they are usually presented by some ordering criterion such as date.

The approach taken by patent searchers contrasts with many other areas of IR research where queries are submitted as unstructured lists of words and best-match models are used to rank the documents. Such best-match models have the advantage that they can take into account not only the presence or absence of a query term in the document, but also its frequency. It is perhaps for this reason that best-match models have found favor within the IR community as this often results in significantly better retrieval performance (in terms of precision and recall). Nevertheless, the Boolean exact-match model remains popular in the patent domain and this is partly due to nature of searches that take place [16, 22]. The usage of the exact-match models stems from the following reasons:

**Custom and Practice:** Practitioners have been trained and are used to formulating Boolean queries. The habit of always performing such queries may make them less likely to change, especially if there current practice is effective in finding the required documents.

**Extra Information Content:** For a given number of query terms, the addition of Boolean operators and brackets adds more information to the queries. Thus very precise queries can be formulated which have a clear interpretation[1].

**Demonstrating Due Diligence:** The fact that there is a crisp cut-off means that a patent searcher does not have to make an arbitrary decision where to stop examining the documents. This protects him/her against the accusation that 'if they had only looked a little further they would have found the document in question.'

**Model Intuitiveness:** Extending a query using either the AND or NOT operator will retrieve fewer documents. This contrasts with best-match models where adding an extra query term will retrieve the same number or more documents. Patent searchers carefully fashion a query specifically to reduce

---

[1] Paradoxically the output of such a model is either 1 or 0 and this contains less information than the real number yielded by best-match models.

the number of retrieved documents to make the exhaustive viewing of each document feasible. However, the NOT operator is known to be problematic since it may lead to the exclusion of a document which was relevant but addressed other topics as well.

Therefore any new model for patent search need to provide the same functionality of exact-match models, or at least handle AND and OR operators, but preferably also the NOT operator. And also, it must provide a crisp cut-off between the retrieved and non-retrieved. In this work, we examine hybrid models which combine theses features of exact-match models with features that improve the access within best match models, in an attempt to obtain the best of both worlds (i.e. exact-match models with improved access).

## 3    Retrievability

The accessibility of information in a collection given a retrieval system has been considered from two points of view, the system side i.e. *retrievability* [8] and the user side *findability* [17]. Retrievability measures provide an indication of how easily a document could be retrieved using a given IR system, while findability measures provide an indication of how easily a document can be found by a user with the IR system. Here we consider the access based measure of retrievability (see [8] for more details and [4, 9–11] for examples of its usage in practice.)

### 3.1    Definition of Retrievability

The general formula for the retrievability measure of a single document given in [8] (with modified notation) is:

$$R(d) = \sum_{q \in Q} p(q) \cdot f(\delta(q, d), \theta) \tag{1}$$

where $Q$ is the set of all possible queries, $p(q)$ denotes the probability of query $q$ being used, $f(\delta(q, d), \theta)$ is the utility function (note that a high value is good) with $\theta$ as a parameter and $\delta(q, d)$ is some measure of the cost involved in accessing $d$ given $q$ (i.e. going down the through the ranked list of documents incurs a cost). It is not possible to create an exhaustive list of queries, so a subset $Q' \subset Q$ is created to form an estimate. Since this subset is usually generated artificially and the queries are not based on any actual information need, we are not able to assign a likelihood to each query other than by assuming $p(q) = \frac{1}{|Q'|}$ is the same for all queries. This then becomes multiplication by a constant. Since we are only interested in this measure relative to other documents, it can be ignored. Thus to provide an estimate $\hat{R}(d)$ of document retrievability we write:

$$\hat{R}(d) = \sum_{q \in Q'} f(\delta(q, d), \theta) \tag{2}$$

To represent the diversity of possible queries that a user could submit, any empirical study will require a very large sample of such queries. It is for this reason that the queries are generated automatically (see Section 4 for details).

## 3.2   Utility Functions

It is assumed that when presented with an ordered list of retrieved documents, the user will start at the top and work their way down. Therefore within IR, the measure of distance $\delta(q, d)$ is the rank of the document. Given that a patent searcher will choose how many documents to examine, we shall use a *cumulative* measure of retrievability where the utility function gives a score of 1 to the top $\tau$ ranked documents, and zero otherwise. In our experiments we shall take measurements of retrievability at five cut offs, where $\tau = 10, 20, 50, 100$ and $200$. Note that, on average, a patent searcher examines around 100-200 documents per query [16, 22].

## 3.3   A Measure of Collection Access

Once we have estimated the retrievability of all the documents in a collection, we wish to calculate some overall measure of access for the collection given the IR system and corpora. Since, the retrieval of one document at a particular rank, means that another document can not be retrieved at the same rank, then documents compete to be retrieved, i.e. if one document appears in the top ten retrieved items then, by definition it will displace another document. Indeed, documents become less retrievable precisely because others become more retrievable. What is of interest here is the distribution of retrievability over the population of documents and whether the retrieval system provides a similar amount of retrievability to each document in the collection (or provide a degree of equality to all documents). For example, we can imagine that we have a retrieval system which only retrieves one particular document regardless of the query. This document would have a very high retrievability score, but all the other documents would have zero retrievability. Since we would like to ensure that patent searchers can access all documents as easily as possible, then making all documents as equally retrievable as possible would improve access to all parts of the collection. Essentially, we would like the retrieval system to afford all documents with similar retrievability. However, due to the characteristics of the documents this might not always be possible - though the aim is to strive for equality.

In economics, there is a standard method of measuring wealth and income distributions to determine the level of equality. Here, we apply this method to the measures of document retrievability. The Lorenz curve [14] provides a graphical representation of the distribution of individual retrievability scores. The documents are ordered by increasing value of their respective score and then the cumulative score is calculated. This is plotted against the cumulative number of documents. Figure 1 gives an example; note that both axes have been normalized. The unbroken line shows complete equality – i.e. all documents are equally retrievable. The heavy, dashed line shows total inequality, only one document is ever retrieved no matter what the query was. The dotted line shows the case where the retrievability scores are uniformly distributed – this data is simulated randomly here.

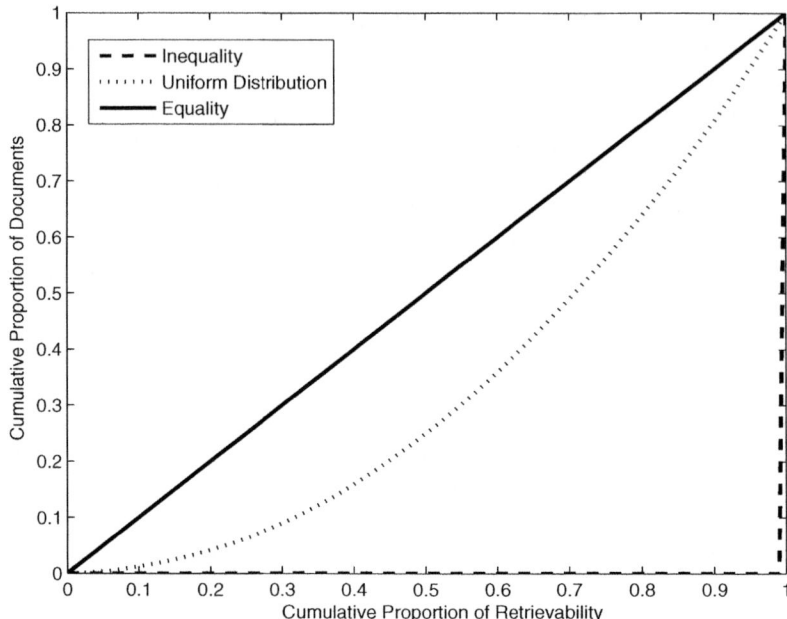

**Fig. 1.** An example of a Lorenz Curve

The Gini [14] coefficient gives a score from zero to 1 indicating the degree of inequality. It is calculated as the area between the 45° unbroken line and the Lorenz curve divided by the total area underneath the 45° line. A score of zero implies total equality; for total inequality, where only one document is ever retrieved, the coefficient approaches 1. In Figure 1, the Gini coefficient for the unbroken line is zero. For the heavy, dashed line it is close to 1 and for lighter dashed line is 0.329. If we assume that the documents have been placed in order of non-decreasing retrievability according to some estimated measure $\hat{R}(d_i)$ and that there are $N$ documents in the collection we can define more formally the Gini coefficient as:

$$1 - \frac{2}{N-1}\left(N - \frac{\sum_{i=1}^{N} i \cdot \hat{R}(d_i)}{\sum_{i=1}^{N} \hat{R}(d_i)}\right)$$

for discrete data, where the documents are indexed by non-decreasing order of $\hat{R}$.

### 3.4   The Retrievability Experiment

Having described the retrievability and access measures, we now summarize the steps undertaken to perform the retrievability analysis, which follows the methodology described in [8].

1. For each corpus of patent documents, we indexed the collection and removed stop words.
2. A large number of queries were automatically generated from the index (see the next section for the method of query generation).
3. For each IR model, each query was used to retrieve documents from the corpus. In Section 6 we describe the best-match, exact-match and hybrid models used in this study. The results from all queries were stored.
4. For each set of results for each retrieval model, the retrievability measurements were calculated for each document.
5. Given the retrievability scores for all documents the overall access measure was calculated (i.e. Gini coefficient) for each model. Here, we attribute a model with a lower Gini coefficient as providing better access to the collection.

The retrieval experiments and query generation was performed using the Lemur Toolkit 4.10 [1].

## 4   Query Generation

To obtain a reasonably accurate estimate of the retrievability scores a sufficiently large number of queries is required. This means that manual generation of queries is not practical. Thus some algorithm is required. In this paper, we adopt the approach previously taken in [8], where a series of one-word and two-word queries were automatically generated given the document in the collection. However, here, only two-word queries were chosen since one-word queries cannot show the effect of Boolean operators. Although the queries used in patent search are often longer than two words, we choose two-word queries because this means there will be just one Boolean operator and it thus affords a comparison between the use of AND and OR within the various IR models. Also, only using two-word queries makes the computation and estimation of the retrievability scores tractable.

Any set of generated queries must relate to the corpus under analysis (i.e. the queries need to contain terms that are in the documents). It is for this reason that query generation method extracts queries using the corpus index. The *collocation* method used in previous studies [8] identifies common phrases in the text, and assumes that they are likely candidates for queries. A collocation is a pair of words which occur more often than would be expected by chance. Queries were derived from collocations found in the document collection since they would tend to relate to some concept which simulates some user information need. The procedure for generating the queries has four distinct stages:

1. The set of all bigrams (pairs of consecutive words) are extracted from the index of the collection.
2. Those bigrams that occur less than a specified number of times are removed.
3. The adjusted point-wise mutual information measure (APWMI) [18] is calculated for each bigram and this is used to sort them in descending order.

4. The first 200 000 of this list of bigrams are disregarded since they contained a lot of stock phrases that were independent of the technical/topical content (e.g. *diagram shows*). The next $n$ queries were then used for the experiment.

These steps are now explained in more detail.

For the purpose of indexing, only those parts of the patent dealing with technical details were used. Thus the sections which listed names and addresses as well as the copyright notice were omitted since these would give rise to common collocations which would not represent plausible queries. Stopwords and punctuation were also removed, as were terms that contained digits or had fewer than two letters. The text was case lowered but stemming was not applied. For each document, each consecutive pair of remaining words was considered as a bigram. Then, each distinct bigram was recorded and the number of occurrences counted. Only those bigrams which occurred at least 10 times were then considered. It is worth noting here that since the IR models under investigation are assumed to have the same indexer with only the matching function varying, a common set of queries was used for all models. Table 1 gives an examples of 10 consecutive queries, which have been automatically generated.

**Table 1.** Examples of Queries Automatically Generated by the Collocation Method

| |
|---|
| diaminopy rimidin |
| indexobjekte datenbestandes |
| leandro chemdbs |
| dimethanesulphonate myleran |
| laktat malat |
| moxon schweda |
| menziesii mirb |
| finegold angelopoulou |
| oxalkylen mischpolymerisate |
| phenanthrenequinone phenylpropanedione |

### 4.1   How Many Queries?

A question arises as to how many queries should be used for the estimation of retrievability. Too few queries will give an inaccurate result, and in particular will mean that the estimated Gini coefficient will be much higher than the true estimate. Thus, we need to ensure that we use enough queries to obtain a reasonably accurate estimate of the Gini coefficient, even though it will not be exact. And while the estimate will not completely accurate it is asymptotically consistent, meaning that inaccuracy of the measurement will tend decrease as the number of queries increase. Of course, there is a practical limit on the number of queries that can be used since each query requires a retrieval operation. Here we have used up to one million queries depending on the collection. In order to facilitate the comparison of models across different corpora, it is important that the document to query ration (DQR) is held constant so that the estimates of the Gini coefficients will not be out of proportion, i.e. if more queries are used on one corpus than another, than its Gini coefficients will be closer to the exact values, which means it is not possible to compare across corpora. For comparisons to be meaningful, the DQR should be kept constant so that the number of queries used is kept proportional to the number of documents in each corpus. Essentially, if corpus A is twice as large as corpus B then it should have twice the number of queries.

## 5  Features for Improving Access

In previous work [9], we conducted a pilot study to determine the differences in retrievability over a number of popular IR models (best-match and exact-match models), and to ascertain to what extent documents were un-retrievable given these models. However, it soon emerged that there were marked differences in the retrievability of documents between the best-match and exact-match models. This led to the hypothesis that there were certain features of the best-match models, not shared with the exact-match models that increased access to the corpus by making the retrievability of documents more equal.

Three features of the best-match models were identified as potentially affecting retrievability. It should be noted that these features are also likely to improve effectiveness but such an investigation is beyond the scope of this paper. However, it was with effectiveness in mind that Fang et al [13] identified a set of constraints that an IR model, ought to adhere to. The chosen features closely relate to some of these constraints. The three features identified were:

1. Sensitivity to Term Frequency,
2. Length Normalization,
3. Convexity.

Each is now considered in turn.

Whereas the traditional Boolean model considers only if a term is present or absent, best-match models such as BM25 and Tf-Idf take into account the number of occurrences. A higher frequency of a given query term will make the document more relevant. This is codified in the Term Frequency Constraint (TFC1) in [13]. It is also common to take account of how frequently the term appears in the entire collection so that very commonly used vocabulary score less that the rarer words. This is consistent with their Term Discrimination Constraint (TDC) [13].

One consequence of incorporating term frequency into a model is that it will tend to score longer documents higher than shorter documents. Although we can argue that, all other things being equal, a longer document is more likely to be relevant since it will contain more information, there is a tendency to over-score longer documents. Thus many models incorporate some length normalization so that shorter documents are not unduly penalized. This is consistent with the Length Normalization Constraint (LNC1) [13].

Fang et al [13] also attempt to define a 'desirable characteristic' of a retrieval model whereby the combination of two distinct query terms in the document will score higher than just occurrences of one or the other. We formalize this concept as *convexity* and provide the following definition. Let $q$ be a query with two terms $w_1, w_2$. Let $d_1$ and $d_2$ be documents which yield precisely the same ranking score for the query $q$. Assume that $d_1$ contains 2 occurrences of $w_1$ and none of $w_2$. Assume also that $d_2$ is identical to $d_1$ except that the 2 occurrences of $w_1$ are replaced with $w_2$. Now, let us assume a third document $d_3$ which is identical to $d_1$ except that only 1 occurrence of $w_1$ is replaced with $w_2$ so that it now contains one of each query term. An IR model will have convexity if, and only if, it always ranks $d_3$ (strictly) higher than $d_1$ and $d_2$.

One key feature of the traditional Boolean model is that it provides a crisp cutoff since each document will be assigned a score of 0 or 1. Adding sensitivity to term frequency means that the scoring of any document *cannot* have just two values. Nevertheless, a crisp cut-off would be possible if there were some attainable minimum score, say 0, representing the Boolean condition being false and a positive value to represent where it is true.

## 6   Retrieval Models

For the purposes of our empirical study, we considered an exact-match retrieval model (i.e. the traditional Boolean model), two well established best-match models (i.e. Tf-Idf and BM25) and a number of hybrid models, which can accept a Boolean query but take into account some or all of these the features introduced above. In an initial study [9], we considered several other hybrid models, but here we perform this larger study on a subset of the best performing retrieval models (i.e. those models that improved access). An IR system consists of an index and a matching model. Since we use the same index throughout the empirical study, the only variation between systems is the matching model.

### 6.1   Traditional Boolean Models

Given that the simulated queries consist of two terms, both of which are considered desirable in any retrieved document, they may only be combined in two ways, namely with an AND and an OR. As a true exact-match model there is no ordering imposed on the retrieved documents, so two possible methods of sorting the retrieved documents were chosen:

1. Chronological order – earliest document first,
2. Reverse Chronological order – latest document first.

We note here that Boolean AND will possess convexity since it clearly gives a higher ranking when both terms are present.

## 6.2   Tf-Idf

This is really a family of retrieval models where the the acronym stands for Term Frequency - Inverse Document Frequency [20]. We shall use the notation $c(w, d)$ to represent the counting function which yields the number of occurrences of $w$ in a document $d$. The document frequency $df(w)$ is the number of documents in the collection which contain at least one occurrence of $w$. There are several formulations for inverse document frequency (Idf). The one chosen here prevents the Idf ever becoming zero even if a term is present in every document.

$$Idf(w) = \log \left( \frac{N + 1}{df(w)} \right) \tag{3}$$

where $N$ is the number of documents in the collection and $q$ is the query. The ranking function multiplies the occurrence of each query term in the document by the Idf measure.

$$\sum_{w \in q \cap d} idf(w) \cdot c(w, d) \tag{4}$$

This we will refer to as *standard* Tf-Idf and note that it is sensitive to term frequency but possesses neither the convexity property nor length normalization. It is, however, possible to address length bias by using pivoted normalization and define *normalized* Tf-Idf as:

$$\sum_{w \in q \cap d} \frac{Idf(w) \cdot c(w, d)}{(1 - b) + b \cdot \frac{|d|}{avdl}} \tag{5}$$

where $|d|$ is the size of the document and $avdl$ is the average document length. We set the parameter $b$ to 0.75 to be the same as the BM25 model below.

If we consider retrieving all the documents whose matching score is greater than zero, we can see that this has the same effect as the Boolean OR since if any of the terms are present the ranking function will have a positive value. However, the retrieved documents will nevertheless be sorted according to the frequencies of the query terms. If more query terms are present or these terms are rarer in the collection, it will score a document higher.

## 6.3   Okapi BM25

This is also a family of matching models and is often referred to as either Okapi or BM25 . The original formulation [21] is based on a probabilistic model. However, various ad hoc changes have been advocated such as that proposed by Fang et al. [13]. It is also common to ignore the query factor of the original formulation (this can be achieved by allowing one of the parameters to approach infinity). The formulation used is:

$$\sum_{w \in q \cap d} idf(w) \cdot \frac{(k_1) \cdot c(w, d)}{k_1 \cdot \left( (1 - b) + b \cdot \frac{|d|}{avdl} \right) + c(w, d)} \tag{6}$$

Two parameters in the model are set to the following value $b = 0.75$ and $k_1 = 1.2$ as is standard practice. As with Tf-Idf, BM25 has the intrinsic OR property and also takes account of term frequency. Because each additional occurrence of the same term makes a diminishing contribution to the overall score, this function also exhibits convexity. This model has been shown to perform well in TREC evaluations and, in particular, to outperform Tf-Idf.

### 6.4    Filtered Term-Frequency Models

The idea of constructing a best-match model which accepts a Boolean query is not new. Salton et al. [19] attempted this some years ago. However, their proposed model does not give a crisp cut-off which we argued for in Section 2 and thus we seek other solutions. One reason we have proposed for using a Boolean query is that the use of the AND operator can actually prevent some documents being retrieved. We propose one approach to add conjunctivity to the two term-frequency models described above. A second approach, the harmonic model [9] was found to yield almost identical results and so is not included here. Thus we use a Boolean expression to filter the results of best-match function. That is, where the Boolean expression is true, the document is scored according to a best-match model such as Tf-Idf and BM25. Where the condition is false then the document is scored at zero. It is worth noting here that both BM25 and Tf-Idf will yield zero if and only if no query terms appear in the document. This is not true for all best-match models and so this approach is only applicable to certain best-match models.

We assume that there is some Boolean query consisting of a set of query terms with the operators AND and OR. The words are extracted and used to calculate a term-frequency model except that if the Boolean expression is false, the matching value is set to zero. If the Boolean expression contains only the OR operator, the matching value will be zero when the Boolean expression is false anyhow. The filter only cuts when there are AND operators. This approach has been used before by Arampatzis et al. [3] on legal queries. The approach can be generalized to Boolean expressions containing the NOT. Terms that are required to be absent are kept within the Boolean expression and if they were present it would set the matching value to zero. However, such terms would not form part of the list of terms used to calculate the Tf-Idf or BM25 function. There is one theoretical drawback with the filtering approach, which is that the filter introduces a discontinuity in the matching function.

We note here that when used with an AND operator, the filtered models will exhibit convexity, whereas the OR operator will not except of BM25 where convexity is present in both cases.

### 6.5    Summary of Models

Since the generated queries used each have two terms, we can consider there to be either an AND or OR operator between them. We note as stated above that applying a Boolean filter makes no difference for an OR operator. Only when AND is used is the result changed.

Table 2. Models Used in Experiment

|  | No Convexity | Convexity |
|---|---|---|
| **Term Presence** | Chronological OR Reverse Chronological OR | Chronological AND Reverse Chronological Boolean Filter |
| **Term Frequency** | Standard TfIdf | Standard TfIdf with Boolean Filter |
| **Term Frequency with Length Normalization** | Normalized Tf-Idf | BM25 BM25 with Boolean Filter Normalized TfIdf with Boolean Filter |

Table 2 summarizes the ten models used according to the presence or absence of the three features. Note that the models that are insensitive to term frequency are all exact match. We expect retrievability to improve towards the bottom of the table.

# 7 Results

We calculated the Gini coefficients of retrievability for each IR model given each corpora from the four patent offices (European, US, Japanese and World) which together make up the MAREC collection [2]. This allowed both comparison of the models, revealing the ones which can give greater access and also comparison between corpora. For this latter purpose, the number of generated queries has been chosen to keep the document to query ratio constant.

Summary statistics relating to the patents contained in the corpora are shown in Table 3 along with the number of queries used. The European and World patent documents were in English, French and German with many documents containing more than one language. Other corpora were in English only. It should be noted that whereas some documents were the full version of the patent, others, specifically in the Japanese corpus, were short summaries or abstracts. The patent documents contained in the four corpora were semi-structured (in XML format) meaning that it was possible to extract automatically certain sections for indexing and not others.

The index used for retrieval was the same as the one used to generate queries; that is it contained only the technical sections of each patent document. Although it would have been more realistic to have used the entire document, an initial study performed on the World collection showed that the results were very similar. Thus to save the computing time and storage space of creating separate indexes, a single index was used for each corpus. Three languages were used in the whole MAREC collection: English, French and German. Some documents had a mixture of languages. Thus a combined three-language stop word list was used. Stemming was not applied. We now address comparison both of the models and the corpora.

**Table 3.** Summary Statistics for MAREC Collection and Generated Queries

| Corpus | Number of Documents | Number of Queries | Document to Query ration (DQR) |
|---|---|---|---|
| European (EP) | 3,508,686 | 415,054 | 8.454 |
| Japanese (JP) | 8,453,560 | 1,000,000 | 8.454 |
| United States (US) | 5,639,471 | 667,112 | 8.454 |
| World (WO) | 1,784,980 | 211,151 | 8.454 |

| Corpus | Number of Terms | Number of Unique Terms | Mean Document Length |
|---|---|---|---|
| European (EP) | 4,936,283,816 | 11,412,080 | 1,406.88 |
| Japanese (JP) | 735,047,852 | 802,875 | 86.95 |
| United States (US) | 16,107,226,982 | 11,281,670 | 2,856.16 |
| World (WO) | 4,288,769,761 | 15,767,937 | 2,402.70 |

**Table 4.** Gini Coefficients for all Scoring Functions using Collocation-based Queries

| Model | OR Relation | | | | | AND Relation | | | | |
|---|---|---|---|---|---|---|---|---|---|---|
| Number of Top Documents Retrieved ($\tau$) | 10 | 20 | 50 | 100 | 200 | 10 | 20 | 50 | 100 | 200 |
| **European (EP)** | | | | | | | | | | |
| BM25 | 0.815 | 0.778 | 0.742 | 0.722 | 0.705 | 0.821 | 0.788 | 0.762 | 0.751 | 0.746 |
| Tf-Idf Std. | 0.965 | 0.947 | 0.916 | 0.889 | 0.861 | 0.916 | 0.892 | 0.861 | 0.842 | 0.828 |
| Tf-Idf Norm. | 0.946 | 0.919 | 0.876 | 0.840 | 0.806 | 0.876 | 0.846 | 0.817 | 0.802 | 0.792 |
| Boolean Ch. | 0.998 | 0.996 | 0.990 | 0.984 | 0.974 | 0.973 | 0.956 | 0.924 | 0.898 | 0.873 |
| Boolean Rev. | 0.998 | 0.996 | 0.991 | 0.984 | 0.975 | 0.974 | 0.957 | 0.925 | 0.899 | 0.874 |
| **Japanese (JP)** | | | | | | | | | | |
| BM25 | 0.712 | 0.609 | 0.487 | 0.423 | 0.384 | 0.709 | 0.601 | 0.462 | 0.374 | 0.305 |
| Tf-Idf Std. | 0.990 | 0.983 | 0.966 | 0.945 | 0.912 | 0.817 | 0.725 | 0.586 | 0.488 | 0.408 |
| Tf-Idf Norm. | 0.988 | 0.981 | 0.964 | 0.941 | 0.906 | 0.805 | 0.705 | 0.556 | 0.452 | 0.368 |
| Boolean Ch. | 1.000 | 1.000 | 0.999 | 0.998 | 0.996 | 0.963 | 0.933 | 0.863 | 0.788 | 0.696 |
| Boolean Rev | 1.000 | 1.000 | 0.999 | 0.998 | 0.996 | 0.962 | 0.931 | 0.863 | 0.789 | 0.699 |
| **United States (US)** | | | | | | | | | | |
| BM25 | 0.873 | 0.842 | 0.804 | 0.779 | 0.755 | 0.886 | 0.864 | 0.840 | 0.826 | 0.814 |
| Tf-Idf Std. | 0.968 | 0.952 | 0.923 | 0.896 | 0.867 | 0.936 | 0.919 | 0.897 | 0.882 | 0.867 |
| Tf-Idf Norm. | 0.955 | 0.932 | 0.891 | 0.854 | 0.817 | 0.910 | 0.891 | 0.869 | 0.855 | 0.841 |
| Boolean Ch. | 0.996 | 0.993 | 0.986 | 0.979 | 0.969 | 0.970 | 0.953 | 0.927 | 0.907 | 0.886 |
| Boolean Rev | 0.996 | 0.994 | 0.988 | 0.981 | 0.972 | 0.973 | 0.957 | 0.933 | 0.913 | 0.893 |
| **World (WO)** | | | | | | | | | | |
| BM25 | 0.846 | 0.812 | 0.770 | 0.746 | 0.734 | 0.890 | 0.888 | 0.892 | 0.897 | 0.901 |
| Tf-Idf Std. | 0.911 | 0.880 | 0.843 | 0.821 | 0.806 | 0.909 | 0.899 | 0.897 | 0.899 | 0.902 |
| Tf-Idf Norm. | 0.886 | 0.849 | 0.805 | 0.780 | 0.764 | 0.897 | 0.892 | 0.893 | 0.897 | 0.901 |
| Boolean Ch. | 0.979 | 0.966 | 0.938 | 0.905 | 0.865 | 0.909 | 0.898 | 0.896 | 0.898 | 0.902 |
| Boolean Rev | 0.989 | 0.981 | 0.958 | 0.928 | 0.886 | 0.920 | 0.905 | 0.899 | 0.900 | 0.902 |

**Table 5.** Mean Occurrence of Bigrams used in Query Set

| Corpus | Mean Occurrence |
|--------|-----------------|
| European (EP) | 250.23 |
| Japanese (JP) | 117.19 |
| United States US | 185.99 |
| World (WO) | 21.96 |

**Table 6.** Mean Frequency of Retrieval for all Matching Models using Collocation-generated Queries

| Corpus | OR Relation | | | | | AND Relation | | | | |
|--------|------|------|------|------|------|------|------|------|------|------|
| Number of Top Documents Retrieved ($\tau$) | 10 | 20 | 50 | 100 | 200 | 10 | 20 | 50 | 100 | 200 |
| European (EP) | 1.181 | 2.357 | 5.863 | 11.640 | 23.011 | 1.140 | 2.224 | 5.238 | 9.580 | 16.789 |
| Japanese (JP) | 1.183 | 2.366 | 5.913 | 11.823 | 23.634 | 1.179 | 2.337 | 5.648 | 10.745 | 19.835 |
| United States (US) | 1.171 | 2.319 | 5.663 | 11.078 | 21.589 | 1.108 | 2.094 | 4.563 | 7.783 | 12.692 |
| World (WO) | 1.178 | 2.349 | 5.830 | 11.501 | 22.184 | 0.914 | 1.447 | 2.048 | 2.367 | 2.568 |

## 7.1  Comparison of Models

Table 4 gives Gini coefficients for the four corpora for all ten models and five retrievability measures for different values of $\tau$. For the first two corpora (EP, JP) the results are very similar to those in the pilot study [9].

1. The term-frequency sensitive models outperform corresponding models that capture only term presence. Specifically, standard Tf-Idf shows greater retrievability than Boolean OR. Also Tf-Idf with an AND filter outperforms the Boolean AND model.
2. Models with length normalization (BM25 and Normalized Tf-Idf) outperform standard Tf-Idf which is not normalized.
3. Models with convexity outperform the corresponding models without. In particular, standard BM25 outperforms standard Tf-Idf which implies that convexity is important when the OR operator is used; it is always present when we use the AND operator.

Thus when each of the model features is present, it gives a more equitable retrievability. For the US corpus, points 1 and 2 are also demonstrated but the story for point 3 is more ambiguous. For smaller numbers of documents retrieved ($\tau = 10, 20, 50$) the AND operator gives better retrievability than OR, except for BM25 where the OR version has convexity anyway. However, as the number of documents retrieved increases 100 or 200, we observe that for the two Tf-Idf variants sometimes the OR outperforms AND.

For the WO corpus, points 1, 2 and 3 are true for the OR operator only. For the AND operator, the results are curious for two reasons. Firstly, the Gini measure does not increase as $\tau$ increases; indeed it sometimes falls. Secondly,

**Table 7.** Percentage of Documents Retrieved n-times for Each Corpora using $\tau = 50$

| Model | OR Relation | | | | | AND Relation | | | | |
|---|---|---|---|---|---|---|---|---|---|---|
| | How Often Retrieved | | | | | | | | | |
| | 0 | 1 | 2–9 | 10 – 99 | ≥ 100 | 0 | 1 | 2–9 | 10 – | ≥ 100 |
| **European (EP)** | | | | | | | | | | |
| BM25 | 36.68 | 13.67 | 31.13 | 18.33 | 0.18 | 40.45 | 13.75 | 29.73 | 15.88 | 0.20 |
| Tf-Idf Std. | 67.43 | 7.28 | 14.40 | 9.91 | 0.97 | 54.88 | 10.53 | 22.66 | 11.29 | 0.64 |
| Tf-Idf Norm. | 61.80 | 7.57 | 16.61 | 13.25 | 0.77 | 49.96 | 10.64 | 25.10 | 13.89 | 0.41 |
| Boolean Ch. | 85.53 | 6.03 | 5.84 | 2.08 | 0.52 | 62.54 | 12.89 | 16.65 | 7.01 | 0.91 |
| Boolean Rev. | 85.65 | 5.97 | 5.85 | 2.04 | 0.50 | 62.88 | 12.75 | 16.49 | 6.97 | 0.91 |
| **Japanese (JP)** | | | | | | | | | | |
| BM25 | 10.78 | 10.79 | 58.19 | 20.23 | 0.00 | 9.71 | 10.69 | 60.89 | 18.71 | 0.00 |
| Tf-Idf Std. | 78.66 | 6.77 | 8.97 | 4.12 | 1.47 | 19.30 | 13.39 | 47.34 | 19.98 | 0.00 |
| Tf-Idf Norm. | 77.22 | 7.34 | 9.67 | 4.28 | 1.49 | 16.12 | 13.21 | 50.82 | 19.86 | 0.00 |
| Boolean Ch. | 96.93 | 1.17 | 1.13 | 0.52 | 0.24 | 52.18 | 17.23 | 22.39 | 7.64 | 0.56 |
| Boolean Rev. | 96.93 | 1.16 | 1.13 | 0.53 | 0.25 | 46.86 | 17.20 | 25.25 | 9.66 | 1.02 |
| **United States (US)** | | | | | | | | | | |
| BM25 | 41.66 | 15.05 | 28.83 | 14.07 | 0.39 | 48.27 | 16.30 | 24.55 | 10.50 | 0.38 |
| Tf-Idf Std. | 68.38 | 7.27 | 14.23 | 9.24 | 0.89 | 58.26 | 13.68 | 19.69 | 7.71 | 0.64 |
| Tf-Idf Norm. | 65.61 | 7.94 | 16.23 | 9.74 | 0.47 | 53.68 | 14.64 | 21.93 | 9.21 | 0.54 |
| Boolean Ch. | 83.08 | 6.17 | 7.52 | 2.69 | 0.54 | 65.39 | 11.59 | 15.81 | 6.49 | 0.71 |
| Boolean Rev. | 84.05 | 5.94 | 6.97 | 2.52 | 0.51 | 66.59 | 11.44 | 15.13 | 6.13 | 0.71 |
| **World (WO)** | | | | | | | | | | |
| BM25 | 41.43 | 12.92 | 27.46 | 17.97 | 0.22 | 68.66 | 10.74 | 15.86 | 4.59 | 0.14 |
| Tf-Idf Std. | 54.02 | 9.52 | 21.30 | 14.56 | 0.60 | 69.32 | 10.62 | 15.42 | 4.47 | 0.16 |
| Tf-Idf Norm. | 48.70 | 10.55 | 23.52 | 16.89 | 0.34 | 68.87 | 10.69 | 15.74 | 4.56 | 0.15 |
| Boolean Ch. | 71.42 | 7.46 | 12.29 | 7.59 | 1.24 | 69.21 | 10.62 | 15.50 | 4.51 | 0.16 |
| Boolean Rev. | 74.55 | 7.83 | 10.94 | 5.54 | 1.13 | 69.47 | 10.75 | 15.27 | 4.34 | 0.17 |

the Gini measures appear very similar for all models, which is a pattern not shown elsewhere. A further investigation shows that this can be attributed to the queries being somewhat different. Table 5 shows the mean occurrences of each bigram corresponding to a query in each entire corpus. For the WO corpus the bigrams are far less frequent and would therefore would expect to be found in far fewer documents. Of course, each of the terms in any bigram could appear separately in any document but the fact these bigrams were selected by having a high APWMI measure implies that occurrences of the terms separately will also be rare. Thus for many of the queries in the WO corpus, there will be few documents which contain both terms and so few documents will be retrieved for each of these queries. This is confirmed in Table 7, which shows the mean number of times each document is retrieved. Note that this statistic is the same for each of the five matching functions. This raises the question as to whether the WO corpus has radically different properties from the other corpora or whether the collocation method has failed to find plausible queries. However, we leave this for future work.

Table 7 shows the frequency with which documents are retrieved. We note that where the Gini coefficient is very high it corresponds to a large number of documents never being retrieved. One explanation of lack of retrievability consistent with this observation is that some documents are regularly pushed down the ranking by others which also match the query. This would explain the very high numbers of non-retrieved documents when using the Boolean model where earlier or later documents always come before them and and thus more frequently retrieved.

### 7.2   Comparison of Corpora

Considering Table 4 again, we can see a marked difference between the corpora in terms of access. Not only do the Gini measures show different values for the same model but also some models perform better on some collections than others. The JP collection has potentially the greatest access if models with convexity are used, yet the worst level of access when using the Boolean OR. It should be noted that the JP collection has much smaller documents than the other three since it consists mainly of short patent summaries whereas the other corpora contain a mixture of summaries and full patents.

## 8   Conclusion and Further Work

From the retrievability analysis that we have conducted, it appears that the hybrid models offer patent searchers the best of both worlds. On one hand they accept a Boolean query and provide a crisp cutoff, thus fulfilling the requirements explained in Section 2. While, on the other hand, they provide improved access to the documents within the collection over the traditional Boolean model. This is because the hybrid models included the three features: term-frequency sensitivity, length normalization and convexity, which have been shown to improve access across the collection. Our analysis also shows that different models provide greater access depending on the collection, and so the choice of model is dependent on the corpora. This research provides an interesting starting point for the analysis and profiling of collections, such as patent corpora, and determining how easily the documents can be found given a particular system. In the case of patent searching, it is very important that the tools that patent searchers employ enable to them to access all parts of the collection as easily as possible. If parts of the collection are not easily accessible then this could lead to missing relevant documents, and doing so could be quite costly.

In future work, we shall consider other ways to estimate the retrievability of documents, by using more sophisticated querying models, and examine other types of retrieval model features which could be incorporated into hybrid models to further improve access.

## Acknowledgements

This project was supported and partly funded by Matrixware. We would like to thank the Information Retrieval Facility for their computation services. We would also like to thank Tamara Polajnar, Richard Glassey and Desmond Elliott for their helpful comments and suggestions on how to improve this work.

# References

1. The lemur toolkit, http://trec.nist.gov/data.html (Last visited 2010)
2. Matrixware research collection (2010), http://www.ir-facility.org/research/data/matrixware-research-collection
3. Arampatzis, A., Kamps, J., Koolen, M., Nussbaum, N.: Access to legal documents: Exact match, best match and combinations. In: TREC 2007: NIST Special Publication 500-274: The Sixteenth Text Retrieval Conference Proceedings, Gaithersburg, MD, USA. NIST (2007)
4. Azzopardi, L., Bache, R.: On the relationship between effectiveness and accessibility. In: Proceedings of the 33th Annual ACM Conference on Research and Development in Information Retrieval, SIGIR 2010 (to appear, 2010)
5. Azzopardi, L., Vanderbauwhede, W., Joho, H.: A survey of patent analysts' search requirements. In: Proceedings of the 33th Annual ACM Conference on Research and Development in Information Retrieval, SIGIR 2010 (to appear, 2010)
6. Azzopardi, L., Vinay, V.: Accessibility in information retrieval. In: Macdonald, C., Ounis, I., Plachouras, V., Ruthven, I., White, R.W. (eds.) ECIR 2008. LNCS, vol. 4956, pp. 482–489. Springer, Heidelberg (2008)
7. Azzopardi, L., Vinay, V.: Document accessibility: Evaluating the access afforded to a document by the retrieval system. In: Evaluation Workshop at the European Conference in Information Retrieval, Glasgow, UK (March 30-April 3, 2008)
8. Azzopardi, L., Vinay, V.: Evaluation methods for information access tasks. In: CIKM 2008 Proceedings of the 17th ACM International Conference on Information and Knowledge Management, California, US, October 26-30. ACM Press, New York (2008)
9. Bache, R., Azzopardi, L.: Identifying retrievability-improving model features to enhance boolean search for patent retrieval. In: Proceedings of the 1st International Workshop on the Advances in Patent Information Retrieval (2010)
10. Bashir, S., Rauber, A.: Improving retrievability of patents with cluster-based pseudo-relevance feedback documents selection. In: CIKM, pp. 1863–1866 (2009)
11. Bashir, S., Rauber, A.: Improving retrievability of patents in prior-art search. To appear ECIR2010, Milton Keynes, England (2010)
12. Bonino, D., Ciaramella, A., Corno, F.: Review of the state-of-the-art in patent information and forthcoming evolutions in intelligent patent informatics. World Patent Information 32(1), 30–38 (2010)
13. Fang, H., Tao, T., Zhai, C.: A formal study of information retrieval heuristics. In: SIGIR '04: Proceedings of the 27th Annual International ACM SIGIR Conference on Research and Development in Information Retrieval, pp. 49–56. ACM, New York (2004)
14. Gastwirth, J.: The estimation of the lorenz curve and gini index. The Review of Economics and Statistics 54, 306–316 (1972)
15. Hunt, D., Nguyen, L., Rodgers, M.: Patent Searching: Tools and Techniques. John Wiley and Sons, Chichester (2007)
16. Joho, H., Azzopardi, L., Vanderbauwhede, W.: A survey of patent users: An analysis of tasks, behavior, search functionality and system requirements. In: Proceedings of the 3rd Symposium on Information Interaction in Context, IIiX 2010 (to appear, 2010)

17. Ma, H., Chandrasekar, R., Quirk, C., Gupta, A.: Improving search engines using human computation games. In: CIKM '09: Proceeding of the 18th ACM Conference on Information and Knowledge Management, pp. 275–284 (2009)
18. Manning, C., Schütze, H.: Foundations of Statistical Natural Language Processing. MIT Press, Cambridge (1999)
19. Salton, G., Fox, E., Wu, H.: Extended boolean information retrieval. Communications of ACM, 1022–1036 (1983)
20. Spärk Jones, K.: A statistical interpretation of term specificity and its application in retrieval. Journal of Documentation 60(5), 779–840 (2004)
21. Spärk Jones, K., Walker, S., Robertson, S.E.: A probabilistic model of information retrieval: Development and comparative experiments (parts 1 and 2). Information Processing and Management 36(6), 493–502 (2000)
22. Tseng, Y.H., Wu, Y.J.: A study of search tactics for patentability search: a case study on patent engineers. In: PaIR '08: Proceeding of the 1st ACM Workshop on Patent Information Retrieval, pp. 33–36. ACM Press, New York (2008)

# Improving Retrievability and Recall by Automatic Corpus Partitioning

Shariq Bashir and Andreas Rauber

Institute of Software Technology and Interactive Systems
Vienna University of Technology, Austria
http://www.ifs.tuwien.ac.at

**Abstract.** With increasing volumes of data, much effort has been devoted to finding the most suitable answer to an information need. However, in many domains, the question whether any specific information item can be found at all via a reasonable set of queries is essential. This concept of Retrievability of information has evolved into an important evaluation measure of IR systems in recall-oriented application domains. While several studies evaluated retrieval bias in systems, solid validation of the impact of retrieval bias and the development of methods to counter low retrievability of certain document types would be desirable.

This paper provides an in-depth study of retrievability characteristics over queries of different length in a large benchmark corpus, validating previous studies. It analyzes the possibility of automatically categorizing documents into low and high retrievable documents based on document properties rather than complex retrievability analysis. We furthermore show, that this classification can be used to improve overall retrievability of documents by treating these classes as separate document corpora, combining individual retrieval results. Experiments are validated on 1.2 million patents of the TREC Chemical Retrieval Track.

## 1 Introduction

The objective of Information Retrieval (IR) systems is to maximize effectiveness. In order to do so, IR systems attempt to discriminate between relevant and non-relevant documents. For measuring effectiveness, metrics such as Average Precision, Q-measure, Normalized Discounted, Cumulative Gain, Rank-Based Precision, Binary Preference (bref) are used [15]. The main limitation of these is, that they focus almost exclusively on precision, i.e. the fact that the (most) relevant documents are returned on top of a ranked list, as this constitutes the primary criterion of interest in most standard IR settings. With evaluation measures such as recall and $F_\beta$, aspects of the completeness of the result set are being brought into consideration.

Most information retrieval settings, such as web search, are typically precision-oriented, i.e. they focus on retrieving a small number of highly relevant documents. However, in specific domains, such as patent retrieval or law, recall becomes more relevant than precision: in these cases the goal is to find all relevant documents, requiring algorithms to be tuned more towards recall at the

A. Hameurlain et al. (Eds.): TLDKS II, LNCS 6380, pp. 122–140, 2010.
© Springer-Verlag Berlin Heidelberg 2010

cost of precision. This raises important questions with respect to retrievability and search engine bias: depending on how the similarity between a query and documents is measured, certain documents may be more or less retrievable in certain systems, up to some documents not being retrievable at all within common threshold settings. Biases may be oriented towards popularity of documents (increasing weight of references), towards length of documents, favor the use of rare or common words; rely on structural information such as metadata or headings, etc.

This gave rise to a new evaluation measure for retrieval systems, namely *retrievability* [1]. Retrievability measures, in how far a system is able to retrieve at least in principle (via a set of reasonably queries) any document in a given corpus. A number of studies on document corpora of limited size have shown, that different retrieval systems perform differently on this task, i.e. that they exhibit a certain bias towards some type of documents (influenced e.g. by document length, vocabulary richness). It can be shown, that in some cases certain documents cannot be retrieved at all within the set of top-c documents returned for any query (within certain constraints, e.g. up to a certain number of query terms) [1,3]. This is due to the fact, that any retrieval system is inherently biased towards certain document characteristics. Bias to some document characteristics [16] is a concept used in Information Retrieval (IR) to describe the fact that a retrieval system gives preference to certain features of documents when ranking retrieval results. There are several techniques for calculating document relevance with respect to query terms. For example **PageRank** [12] calculates web page relevance by favoring large inlink over small inlink counts. In PageRank-style algorithms, if some documents have a higher number of inlinks, they will be ranked higher in the result list. Therefore, the PageRank algorithm is highly biased towards popular documents. Probabilistic retrieval systems such as **BM25** [14], **tf-idf** and **BM25F** [13] are biased towards documents which contain high term frequencies and many different terms, i.e. they favour long documents. Whatever the relevance criterium used in a retrieval system, the main purpose of introducing relevance in query terms is favouring certain types of documents over others, so that the users can retrieve the most relevant documents quickly from top rank results. There is a severe risk that a certain number of documents cannot be found in the top-n ranked results via any query terms that they would actually be relevant for, which ultimately decrease the usability of the retrieval system [2].

Using retrievability measurement, a document corpus can be analyzed, identifying, which documents are highly retrievable (i.e. they can be found by many queries), and which ones show low, down to no retrievability at all, i.e. they cannot be found in the top-c results by any query under a specific retrieval model. On top of this, research indicates (again on datasets of limited size) that it may be possible to identify for a given retrieval system, which documents are likely to show high or low retrievability based on document characteristics, i.e. without performing extensive retrievability measurement [4].

If the assumptions put forward by these results hold true for larger corpora as well, this raises the question, whether we may be able to improve overall retrievability by treating a document corpus as consisting of two different sub-corpora of documents, namely those with high and low retrievability in a given retrieval system. The experiments presented in this paper show, that this is in fact the case.

However, before validating this hypothesis, it is important to double-question the results reported so far in literature on retrievability analysis. While extensive experiments involving a massive amount of query processing were performed, all numbers published on retrievability results suffer from some limitations, by it either that only a comparatively small number of documents was used (e.g. 7,000 or 50,000 docs in [3,4], and/or that only rather short queries (only up to 2 terms combinations in [1]) were processed, and/or limiting the total number of queries issued per document rather than creating an exhaustive set of queries (e.g. max. 90 queries in [4]). Specifically the latter will usually penalize longer documents that potentially could be found via a larger number of (more exotic) query terms combinations, which would also reflect realistic settings in many search scenarios. This also seems to be reflected in the somewhat counter-intuitive figures published in said research, showing a higher retrievability bias (and thus lower overall retrievability) for longer queries, while intuitively higher retrievability should be expected for more specific (i.e. longer) queries.

In this paper we present a series of experiments on a large-scale document corpus in the same application domain as the previous studies on retrievability, i.e. patent retrieval. A representative benchmark corpus of 1.2 million patents used in the TREC Chemical Retrieval Track (TREC-CRT) is being used to validate the hypothesis of uneven retrievability in a large corpus [11]. Specifically, we verify whether retrievability is lower even when using longer queries. To this end we first perform standard exhaustive retrievability evaluation on short queries, followed by query generation using longer queries for documents exhibiting low retrievability on short queries.

We then replicate experiments to classify the entire document corpus into documents with high and low retrievability, yet using a significantly simpler set of features. Extensive experiments analyze the parameter settings required to obtain a suitable training corpus defining the classes of documents showing extremely high retrievability (i.e. dominating result lists on a huge number of queries), as well as documents showing extremely low retrievability, i.e. which are virtually impossible to retrieve.

Having classified the entire document corpus into these two categories, we then perform retrieval by treating these classes as independent partitions, processing queries independently for each and subsequently combining the result sets. We can show that this helps in increasing overall retrievability, reducing the dominance of certain documents in query processing and thus reducing the bias of a retrieval system. This approach thus provides a higher probability of being able to at least potentially find each document in a corpus.

Finally, an evaluation over the TREC-CRT prior art search task reveals, that this retrieval via two partitions of documents with high and low retrievability also helps in increasing overall recall for all baseline systems evaluated.

The remainder of this paper is structured as follows. Section 2 reviews related work on retrievability analysis, emphasizing the key messages learned from these as well as pointing to some limitations in the evaluations. Section 3 first introduces the concept of retrievability measurement and presents a modified score considering a potential bias introduced by the way queries are generated. It then describes the TREC-CRT benchmark corpus used for the experiments in this paper and retrievability results for several configurations of this corpus and different retrieval models. Section 4 then presents an approach to automatically classify documents into potentially high and low retrieval classes based on features capturing term distribution characteristics. Section 5 finally evaluates retrieval performance on the partitioned corpus, analyzing both retrievability as well as recall for the TREC-CRT prior art task. Section 6 briefly summarizes the key lessons learned and points to future work to evaluate the impact on real-life systems.

## 2 Related Work

### 2.1 Patent Retrieval

Patent Retrieval is a highly recall-oriented domain, aiming at identifying all documents relevant to a particular query. Several specific types of query processes may be identified in this domain, such as

**Priot Art Search:** This is a core step when planning to file a new patent application. Here, a survey is conducted in each national intellectual office for checking whether there exist any inventions similar to a given patent application. The mechanism that is generally used when collecting relevant patents applications for such a survey is keyword based query. Query terms are mostly extracted from the Claim sections. Query expansion is used to add related terms for improving the breath of the retrieval process [8].

**Invalidity Search:** In invalidity search the examiners have to find out the existing patent specifications that describe the same invention for collecting claims to make a particular patent invalid. In this search process, the examiners extract relevant query terms from patent applications particular from the Claim sections for creating query sets [9].

**Right-to-Use:** Right-to-use searches are conducted prior to marketing a new product for confirming whether a new patent application is infringing any existing patent application or not. In this application area, the method that is generally used is keyword based retrieval. However, query terms that are used for searching documents do not depend solely on a single patent application. Clustering is also widely used for identifying more target oriented queries which can cover all the related applications [6].

There are also many other patent processing applications such as patent map generation, current awareness search, legal status report, patenting activity report and trend mining. In all of these applications knowledge discovery methods such as data mining and machine learning are generally used for discovering competitive intelligence, which are not directly related to keyword based search.

## 2.2    Evaluating Retrievability

The evaluation of retrieval systems has always received much attention in the IR research community. Conventionally, retrieval systems are evaluated using a variety of precision and recall based measures [15]. However, these do not evaluate, what we can find and cannot find in a collection. Yet, for some specific retrieval applications like *patents* (or the legal domain in general), recall is considered more important than precision.

In addition to using traditional IR metrics for evaluation, Azzopardi et al. [1] introduce a measure for evaluating systems on the basis of retrievability scores of individual documents. It measures, how likely a document can be found at all by a specific system, with the analysis of the individual retrievability scores of documents performed using Lorenz curves and Gini coefficients. Their experiments with AQUAINT and .GOV datasets reveal that with a TREC-style evaluation a proportion of the documents with very low retrievability scores (sometimes more than 80% of the documents) can be removed without significantly degrading performance. This is because the retrieval systems are unlikely to ever retrieve these documents due to the bias they exhibit over the collection of documents.

In [3] we analyze retrievability of documents specifically with respect to relevant and irrelevant queries to identify, whether highly retrievable documents are really highly retrievable, or whether they are simply more accessible from many irrelevant queries rather than from relevant queries. That evaluation was based on using a rather limited set of queries. Experiments revealed, that 90% of patents which are highly retrievable across all types of queries, are not highly retrievable on their relevant query sets.

One of the limitations of the approaches published so far is, that – due to the enormous amount of queries involved – only short (max 2-terms) queries are usually evaluated. Where longer queries are selected, they are limited to a drastically small subsets of all potential longer queries, virtually eliminating their effect on retrievability analysis for a document. We are thus trying to replicate a few of the experiment set-ups to study, in how far more specific, longer queries can help mitigate the problem of low retrievability. Jordan et al. [10] consider controlled query generation for evaluating the impact of retrieval systems performance. The main purpose of their study was to expose the performance of different algorithms, specifically how they react to queries of varying length and term quality (in case of noisy terms), which may also have an impact on retrievability.

Another caveat may lie in the retrievability measure used, which does not consider the (different) numbers of queries generated for each document. Both approaches rely on exhaustive query generation based on terms combinations of a document's vocabulary. This leads to drastically different numbers of queries that can retrieve particularly vocabulary-rich, longer documents. We thus propose a slight adoption of the retrievability score considering the number of queries used to try to retrieve a particular document.

Techniques that address query expansions for the legal domain have been proposed by Custis et al. [5]. They evaluate query expansion methods for legal domain applications retrieval on the basis of query document term mismatch. For this purpose, they systematically introduce query document term mismatch into a corpus in a controlled manner and then measure the performance of IR systems as the degree of term mismatch changes.

For realistic evaluations of patent retrievability, the generic query generation approaches will need to be fine-tuned to specific domains. Fujii [7], for example, applies link analysis techniques to the citation structure for efficient patent retrieval. They first perform text based retrieval for obtaining the top-$c$ patents. They then compute citation scores based on PageRank and topic-sensitive citation-based methods. Finally, both the text-based and citation-based scores are combined for better ranking. Also, more recently, full patents start being used as a query instead of selecting relevant keywords from them for prior-art search [18].

Applying these concepts may lead to more realistic retrievability analysis results, once the principles and limitations of the new measure are fully understood.

## 3    Retrievability Evaluation

### 3.1    Relative Retrievability Measurement

Given a retrieval system $RS$ with a collection of documents $D$, the concept of retrievability is to measure how well each document $d \in D$ is retrievable within the top-$c$ rank results of all queries, if $RS$ is presented with a large set of queries $q \in Q$. Retrievability of a document is essentially a cumulative score that is proportional to the number of times the document can be retrieved within that cut-off $c$ over the set $Q$ [1]. A retrieval system is called best retrievable, if each document $d$ has nearly the same retrievability score, i.e. is equally likely to be found. More formally, retrievability $r(d)$ of $d \in D$ can be defined as follows.

$$r(d) = \sum_{q \in Q} f(k_{dq}, c) \tag{1}$$

$f(k_{dq}, c)$ is a generalized utility/cost function, where $k_{dq}$ is the rank of $d$ in the result set of query $q$, $c$ denotes the maximum rank that a user is willing to proceed down the ranked list. The function $f(k_{dq}, c)$ returns a value of 1 if $k_{dq} \leq c$, and 0 otherwise.

Retrievability inequality can further be analyzed using the Lorenz Curve. Documents are sorted according to their retrievability score in ascending order, plotting a cumulative score distribution. If the retrievability of documents is distributed equally, then the Lorenz Curve will be linear. The more skewed the

**Table 1.** Experiment set-up: number of queries generated for each subset of documents (without duplicates)

| Experiment Method | queries | no docs | 1-terms | 2-terms | 3-terms | 4-terms |
|---|---|---|---|---|---|---|
| A = complete | all-queries | 1.2 million | 437,038 | 236,513,386 | - | - |
| B = 20 doc top/bottom | all queries | 20 | 8,568 | 1,239,449 | 52,668,550 | - |
| C = random 5% (training) | 20% cut | 60,000 | 46,693 | 25,672,536 | 1,689,346,590 | 3,635,659,348 |
| D = prior-art | 20% cut | 34,200 | 29,347 | 10,926,552 | 1,037,061,490 | 2,366,376,269 |

curve, the greater the amount of inequality or bias within the retrieval system. The Gini coefficient $G$ is used to summarize the amount of bias in the Lorenz Curve, and is computed as follows.

$$G = \frac{\sum_{i=1}^{n}(2 \cdot i - n - 1) \cdot r(d_i)}{(n-1)\sum_{j=1}^{n} r(d_j)} \qquad (2)$$

where $n = |D|$ is the number of documents in the collection sorted by $r(d)$. If $G = 0$, then no bias is present because all documents are equally retrievable. If $G = 1$, then only one document is retrievable and all other documents have $r(d) = 0$. By comparing the Gini coefficients of different retrieval methods, we can analyze the retrievability bias imposed by the underlying retrieval system on the given document collection.

However, the retrievability measure as defined above is a cumulative score over all queries. Thus, longer documents that contain a larger vocabulary, potentially have a higher retrievability score than shorter documents. While this is desirable as a general measure of retrievability, in settings where the actual set of queries is created directly from the documents to be found, this may have a negative impact. This is because a larger number of queries are generated for these longer documents. We thus propose for evaluations like these to normalize the cumulative retrievability score by the number of queries that were created from and thus potentially can retrieve a particular document.

$$r(d) = \frac{\sum_{q \in Q} f(k_{dq}, c)}{|\hat{Q}|} \qquad (3)$$

where $\hat{Q}$ is the set of queries that can retrieve $d$ when not considering any rank cut-off factor.

## 3.2   Experiment Set-Up

We use the 1.2 million patents from the TREC Chemical Retrieval Track (TREC-CRT)[1], allowing validation of retrievability measurements on a large-sale corpus within a recall-oriented domain [11].

Retrievability is evaluated for three different retrieval models, namely standard TFIDF based retrieval ranking by the sum of *tfidf* values for the query

---

[1] Available at http://www.ir-facility.org/research/evaluation/trec-chem-09

**Table 2.** Number of queries generated and average number of documents retrieved per query (Experiment D)

| Query Set | # Queries | avg. #docs retrieved |
|-----------|-----------|----------------------|
| Single Term Queries | 29,347 | 33,542.94 |
| Two Terms Queries | 10,926,552 | 21,675.46 |
| Three Terms Queries | 1,037,061,490 | 12,178.82 |
| Four Terms Queries | 2,366,377,269 | 6,472.35 |

terms; the OKAPI retrieval function (BM25) [14]; and a Language Modeling approach based on Dirichlet Smoothing (LM) [19].

For each document in the corpus a set of queries is generated using all terms that appear more than once in the document. In total, four subsets of queries are created, consisting of all single terms, 2-, 3-, and 4-terms combinations. These queries are then posed against the complete corpus of 1.2 million documents as boolean queries with subsequent ranking according to the chosen retrieval model to determine retrievability scores as defined in Equ. 3.

An overview of the various retrievability experiments performed is provided in Table 1. First, initial retrievability scores are determined using the complete set of queries for single term and 2-terms queries, resulting in sorting of documents according to their cumulative retrievability score (experiment A).

As terms distributions in a document tend to have an impact on retrievability, we then selected the 10 documents with the highest and lowest average terms frequencies. For these 20 documents, an exhaustive set of 3-terms and 4-terms queries is passed against the entire corpus to determine, whether low retrievability also persists over longer queries, or whether previously reported results along this line could be an artifact of processing only a rather small subset of the entire set of longer queries (experiment B).

After validating that, indeed, documents showing high/low retrievability over one and two terms queries also show corresponding retrievability over longer queries, and validating that this behavior is not impacted by selecting only a subset of queries, a random 5% subset of the entire corpus was selected (experiment C). Exhaustive sets of single term as well as 2-terms queries were combined with a set of 3- and 4-terms queries with a limit of 20% of the maximum number of potential queries (i.e. terms permutations)

A last group of retrievability experiments are performed using the set of 34,200 patents that are used as ground-truth in the evaluation of the 1,000 *prior art search task* of TREC-CRT, where they are referenced as relevant prior art patents [11]. This subset was chosen to evaluate, whether improving retrievability also improves accuracy for recall-oriented retrieval results, or whether promoting low-retrieval documents would harm accuracy. It allows us to verify whether predominantly documents with a low retrievability score are missed in the prior art retrieval process.

For this subset, an exhaustive set of queries for single term and 2-terms queries, as well as 20% of all 3- and 4-terms queries have been processed against

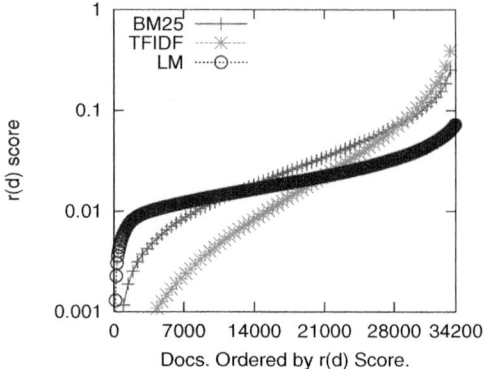

**Fig. 1.** Retrievability inequality for 3 retrieval systems, corpus (Experiment) A, rank cut-off c=150

the entire corpus of 1.2 million documents (experiment D). Each of these experiments was performed for the three retrieval models.

The exact configuration of the document corpora (document IDs) are available from our webpage[2]. Table 2 lists the number of queries generated, for example, for experiment corpus D and the average number of documents returned per query. This provides an indication of the impact for the rank cut-off factor. It also shows the significant decrease of the number of documents returned for longer and thus more specific queries – which, however, have to be selected from a drastically increasing number of potential query terms combinations.

Figure 1 compares the retrievability inequality of the 3 different retrieval models by sorting documents according to their $r(d)$ score. Ideally, all documents would be equally retrievable, i.e. they can be found by an equal fraction of all queries, resulting in a horizontal line. In reality, a significant number of documents (a few hundred for BM25, several 1,000 for TFIDF) cannot be found via any query, via a small number of documents are returned for a huge number of queries. TFIDF shows a stronger bias than BM25, whereas LM shows the lowest bias of all systems, not having any unretrievable documents.

Detailed evaluation of the retrievability values of the 20 documents in corpus (Experiment) B revealed, that the retrievability values determined on 2-terms and 3-terms queries lead to almost identical rankings by r(d) score. Thus, using r(d) scores based on the aggregated results of exhaustive single and 2-terms queries, combined with a sufficiently large number of 3- and 4-terms queries seems to be a solid basis for further analysis. However, better results may be obtainable by using more sophisticated query generation strategies which create the most probable subset of queries according to real-life scenario assumptions.

---

[2] url.witheld.for.review/page/not/publicly/linked.xxxx

# 4   Partitioning the Corpus

The experiments above reveal that, indeed, a corpus consists of documents that show highly different behavior in retrievability. Some documents are returned within the top-$c$ results for a huge number of queries, possibly suppressing others that almost never show up within the top-$c$ results for any query. This means that these documents are virtually inexistent for a searcher. One of the goals in recall-oriented application domains is to esure that all relevant documents are potentially found. We thus need to devise ways to ensure that documents exhibiting low retrievability can also be retrieved by queries that they are potentially relevant for. In order to do so, we propose to split a document corpus into two categories, consisting of documents with high and low retrievability, respectively. These two partitions can subsequently be accessed seperately, possibly via retrieval models that are optimized for the document characteristics, ensuring better overall retrievability.

The obvious way to divide a corpus into documents with high and low retrievability by performing extensive retrievability analysis unfortunately is prohibitive for any realistically-sized corpus. We thus pick up an idea proposed in [4] and try to classify documents into these two categories via a set of surface-level features. While the authors in that study propose a rather complex and extensive set of features to describe documents via co-location of word pairs within certain windows, we apply a simpler set of features that seems to compare favorably with the published results (although they cannot be compared directly as a different document corpus is used).

## 4.1   Features for Retrievability Classification

We compute a number of statistical and information-theoretic features from these documents, resulting in an only 10-dimensional feature vector, capturing the distributional characteristics of terms within a document and over the whole corpus.

- **Normalized Average Term Frequencies (NATF):** average of normalized term frequencies of all terms in a document, calculated as

$$NATF = \frac{\sum_{t \in T_d} \frac{f_{(t,d)}}{|d|}}{|T_d|} \qquad (4)$$

  where $T_d$ represents the set of all unique terms in a document $d$. $f_{(t,d)}$ is the frequency of term $t$ in $d$ (also referred to as $tf$), and $|d|$ represents the length of document.
- **Number of Frequent Terms (freq):** calculates, how many terms have a term frequency larger than a pre-defined threshold of, in our case, 6. This allows us to capture uneven distributions, identifying the absolute number of frequent terms, especially in collections with documents of drastically different length.

- **NATF of Frequent Terms (NATF_freq):** calculates NATF only for frequent terms, i.e. terms having a term frequency larger than e.g 6, to eliminate the impact of a potentially large number of rare terms having only low $tf$ values.
- **Gini Coefficient of Term Frequencies (GC_terms):** measures how balanced the distribution of term frequencies is within a document. Similar to evaluating retrievability inequality, GC_terms captures, whether all terms have similar or rather different $tf$ values.

$$GC\_terms = \frac{\sum_{t \in T_d} (2 \cdot i(t) - |T_d| - 1) \cdot f_{(t,d)}}{(|T_d| - 1) \sum_{t \in T_d} f_{(t,d)}} \tag{5}$$

where $i(t)$ is the index of term $t$ in set $T_d$ after sorting terms in ascending order of their frequencies.
- **Number of Frequent Terms based on Gini-Coefficient (freq_GC):** rather than using a fixed threshold as for NATF_freq, terms with the highest $tf$ values are iteratively removed until the resulting Gini Coefficient for the entire document drops below 0.25, i.e. is rather homogeneous. The number of terms removed provides a different measure for the number of frequent terms contributing to retrieval inequality.
- **Average Document Frequency (ADF):** measures in how far a document consists of rather common or rather specialized vocabulary by summing up the df values of its vocabulary.

$$ADF = \frac{\sum_{t \in T_d} f(d, t)}{|T_d|} \tag{6}$$

where $f(d, t)$ represents the document frequency $(df)$ of term t in D.
- **Frequent Terms with Low Document Frequency (freq_low_df):** measures, how many frequent terms in a document have a rather low document frequency, i.e. are exotic terms in the corpus. Thresholds are set to min $tf$ of 6, and max $df$ of 3,000.
- **Average Document Frequency of Frequent Terms (ADF_freq):** captures the exoticity of the frequent terms in the vocabulary of a document.
- **Relative Term Frequency (TF_rel):** Relative term frequency captures, how the term frequencies in the current document compare to the term frequencies of the subset of the documents where the respective terms have the highest term frequencies. It is calculated by determining the average top term frequency (ATTF) of each term in the top 10% of documents where this term has the highest term frequency.

$$ATTF(t) = \frac{\sum_{d \in \hat{D}_t} f(t, d)}{|\hat{D}_t|} \tag{7}$$

where $\hat{D}_t$ is the 10% set of documents that have the highest $tf$ value for term $t$. These values are subsequently aggregated for a document as the relation between the ATTF and the $tf$ value in the given document.

$$TF\_rel = \frac{\sum_{t \in T_d} \frac{ATTF(t)}{f(t,d)}}{|T_d|} \tag{8}$$

**Table 3.** Classification accuracy (high/low retrievable), Naïve Bayes

| Retr. | rank cut-off factors | | | | |
|---|---|---|---|---|---|
| Sys. | 50 | 100 | 150 | 250 | 350 |
| **TFIDF** | 85% | 84% | 83% | 82% | 79% |
| **BM25** | 82% | 81% | 81% | 80% | 77% |
| **LM** | 80% | 79% | 78% | 77% | 76% |

**Table 4.** Percentage of bottom and top r(d) score documents used as basis for training set definition (tr. low and tr. high)and resulting classification of entire corpus into low and highly retrieval documents.

| Retr. System | tr. low | tr. high | % clas. high | % class. low |
|---|---|---|---|---|
| **TFIDF** | 35% | 35% | 56% | 44% |
| **BM25** | 25% | 40% | 45% | 55% |
| **LM** | 25% | 40% | 42% | 58% |

If TF_rel is high, then the document has rather low $tf$ values for its terms compared to the top $tf$ values for these terms in the corpus.

– **Patent Length (PL):** represents the length of document in words ($|d|$).

## 4.2   Model Training

In order to train a model, we first need to identify a suitable training set configuration, picking certain subsets from the documents showing low and high retrievability. We perform retrievability analysis on corpus (Experiment) C, i.e. a random 5% selection of all documents of the TREC-CRT corpus for which retrievability scores are calculated with exhaustive 1- and 2-terms queries, as well as 20% of all 3- and 4-terms queries, with a rank cut-off factor of $c = 150$.

The set of documents is subsequently split into 3 subsets consisting of documents with high, medium and low retrievability scores. Training instances are picked only from the set of documents with high and low retrievability. A number of different configurations have been evaluated using 10-fold cross-validation training a Naïve Bayes classifier as implemented in the WEKA toolkit [17] to determine the optimal split of the training corpus. Experiments setting the split in 5% increments for documents with low and high $r(d)$ as the bottom and top 5% to 60% revealed that the optimal split was to use the bottom 25% and the top 40% documents in the case of the BM25, whereas the optimal split for the TFIDF retrieval model was both a top and bottom threshold of 35%.

Figures 2 and 3 show the resulting classification accuracies with either the top or bottom threshold fixed, varying the other. The percentage of documents classified into the respective classes when applying this classifier to the entire document collection is also depicted in these figures. We furthermore analyzed

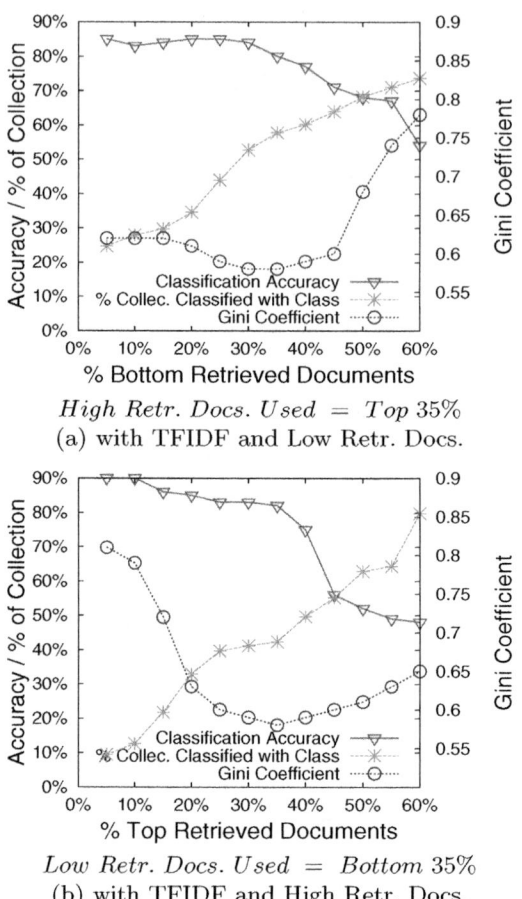

*High Retr. Docs. Used = Top* 35%
(a) with TFIDF and Low Retr. Docs.

*Low Retr. Docs. Used = Bottom* 35%
(b) with TFIDF and High Retr. Docs.

**Fig. 2.** Impact of selecting training documents from different subsets of the training corpus. rank cut-off c=150, TFIDF model, (a) fixing upper threshold 35%, (b) fixing lower threshold at 35%.

the classification accuracy on the training set for different rank cut-off factors as summarized in Table 3.

Details on the final classifier training split are provided in Table 4. For the TFIDF retrieval model, 56% of all documents were classified as potentially having low retrievability with a split of 35/35 for top/bottom categories for training set determination from which the actual training documents were sampled via 10-fold cross-validation. For BM25, the optimal split was 25/40, resulting in 45% of the entire corpus being classified as having potentially low $r(d)$ scores, whereas for the LM approach 42% were classified as low-retrievable with a 25/40 training set delimiter.

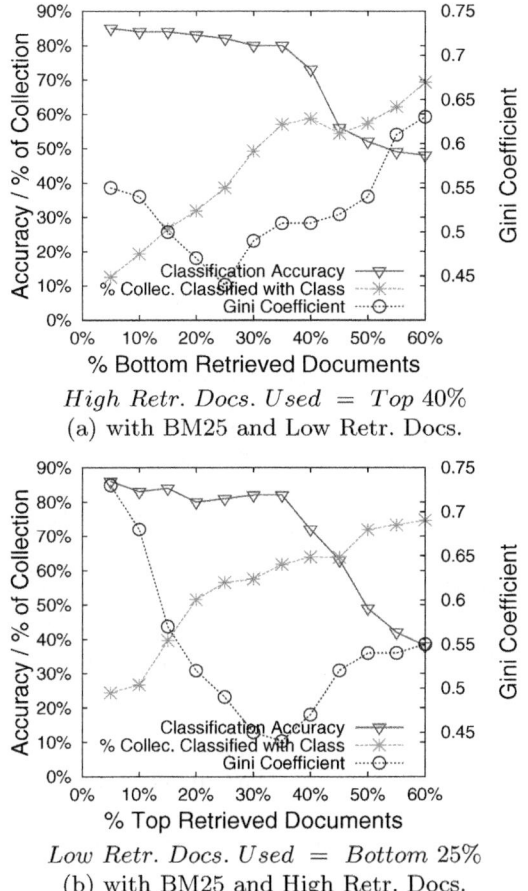

*High Retr. Docs. Used = Top* 40%
(a) with BM25 and Low Retr. Docs.

*Low Retr. Docs. Used = Bottom* 25%
(b) with BM25 and High Retr. Docs.

**Fig. 3.** Impact of selecting training documents from different subsets of the training corpus. rank cut-off `c=150`, BM25 model, (a) fixing upper threshold 40%, (b) fixing lower threshold at 25%.

## 5   Partition-Based Retrieval

Once a document corpus has been divided into two partitions containing documents with potentially high and low retrievability scores, queries can be passed to these corpora independently. On each partition queries are processed independently, and the final result set is merged to form a single result set. This ensures that the final result set will always include also documents having a low retrievability score, i.e. that would rarely or never have been returned within a certain rank cut-off in a standard retrieval setting independent of the query.

A number of merging principles can be envisaged. While relative similarity scores may be used, we used only rather simpler merge strategies, namely

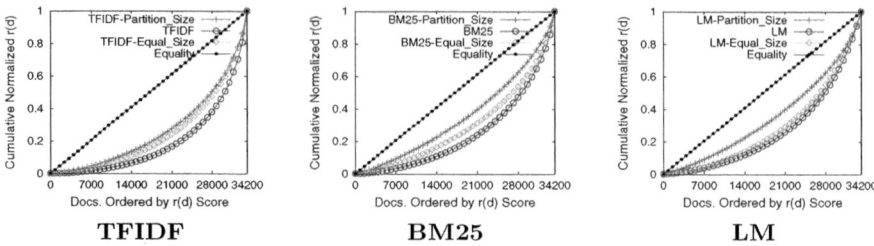

TFIDF    BM25    LM

**Fig. 4.** Visualization of Documents Retrievability using Lorenz Curve, rank cut-off (c=150). Equality refers to an optimal system which has no bias.

**Table 5.** Bias of IR systems with different rank cut-off (c) factors and query sets

| Retr. Sys. | Approach | Four Terms Queries | | | | | Three Terms Queries | | | | | Two Terms Queries | | | | |
|---|---|---|---|---|---|---|---|---|---|---|---|---|---|---|---|---|
| | | 50 | 100 | 150 | 250 | 350 | 50 | 100 | 150 | 250 | 350 | 50 | 100 | 150 | 250 | 350 |
| BM25 | *no split* | 0.75 | 0.67 | 0.56 | 0.54 | 0.51 | 0.63 | 0.56 | 0.52 | 0.51 | 0.49 | 0.50 | 0.49 | 0.44 | 0.41 | 0.37 |
| | *Part_Size* | 0.54 | 0.50 | 0.42 | 0.40 | 0.36 | 0.46 | 0.39 | 0.36 | 0.35 | 0.33 | 0.42 | 0.35 | 0.31 | 0.28 | 0.24 |
| | *Equal_Size* | 0.63 | 0.60 | 0.51 | 0.50 | 0.44 | 0.5 6 | 0.47 | 0.45 | 0.43 | 0.42 | 0.49 | 0.44 | 0.37 | 0.35 | 0.30 |
| TFIDF | *no split* | 0.88 | 0.83 | 0.72 | 0.65 | 0.64 | 0.77 | 0.70 | 0.61 | 0.57 | 0.56 | 0.69 | 0.62 | 0.53 | 0.49 | 0.44 |
| | *Part_Size* | 0.70 | 0.67 | 0.59 | 0.51 | 0.49 | 0.59 | 0.52 | 0.46 | 0.41 | 0.39 | 0.49 | 0.44 | 0.37 | 0.33 | 0.29 |
| | *Equal_Size* | 0.78 | 0.74 | 0.66 | 0.58 | 0.56 | 0.67 | 0.59 | 0.52 | 0.47 | 0.44 | 0.58 | 0.56 | 0.46 | 0.41 | 0.34 |
| LM | *no split* | 0.64 | 0.54 | 0.43 | 0.40 | 0.39 | 0.60 | 0.53 | 0.48 | 0.46 | 0.43 | 0.54 | 0.46 | 0.40 | 0.37 | 0.33 |
| | *Part_Size* | 0.44 | 0.38 | 0.29 | 0.25 | 0.22 | 0.40 | 0.34 | 0.30 | 0.28 | 0.26 | 0.38 | 0.34 | 0.27 | 0.25 | 0.23 |
| | *Equal_Size* | 0.52 | 0.47 | 0.38 | 0.34 | 0.32 | 0.49 | 0.43 | 0.41 | 0.39 | 0.36 | 0.45 | 0.39 | 0.34 | 0.30 | 0.28 |

- **equal_size**: an equal number of documents from the low and high retrievable subsets are returned, i.e. for a given cut-off factor $c$, $c/2$ documents were taken from each partition
- **partition_size**: in this case, the number of documents included in the final result set is relative to the size of the two partitions.

## 5.1   Retrievability Analysis

Figures 2 and 3 show the Gini coefficients using equal size based merging for the splits determined on the training set as an overlay to the parameter estimation process. These basically reveal, that optimal retrievability (i.e. lowest Gini coefficient) nicely co-insides with the configuration of training set thresholds that lead to the highest accuracy in the subsequent classification model.

Figure 4 shows the retrievability inequality of different IR systems using Lorenz Curves with a rank cut-off factor $c = 150$ with two different merging strategies in comparison to the default retrieval setting using a single corpus *(with all queries)*. It clearly shows that the retrieval inequality is lower when using the split corpus approach for all retrieval models. Merging the result set from the two partitions based on their relative size consistently leads to the lowest bias.

**Table 6.** Recall of IR systems with R150 for TREC-CRT prior-art search task on partitioned corpus and without partitioned corpus

| Retr. Sys. | Approach | rank cut-off factor | | | |
|---|---|---|---|---|---|
| | | R150 | R350 | R550 | R750 |
| **TFIDF** | Without Split | 0.008 | 0.014 | 0.021 | 0.028 |
| | Partition_Size | 0.024 | 0.057 | 0.080 | 0.102 |
| | Equal_Size | 0.014 | 0.029 | 0.048 | 0.063 |
| **BM25** | Without Split | 0.022 | 0.042 | 0.055 | 0.076 |
| | Partition_Size | 0.077 | 0.138 | 0.177 | 0.216 |
| | Equal_Size | 0.039 | 0.084 | 0.115 | 0.168 |
| **LM** | Without Split | 0.021 | 0.042 | 0.061 | 0.080 |
| | Partition_Size | 0.074 | 0.116 | 0.184 | 0.242 |
| | Equal_Size | 0.043 | 0.083 | 0.147 | 0.177 |

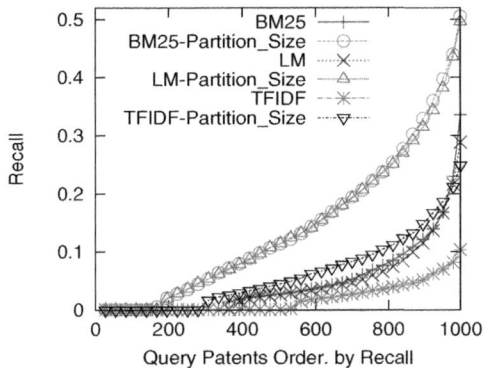

**Fig. 5.** Recall of IR systems for TREC-CRT prior-art search task on partitioned corpus and without partitioned corpus. Query Patents are ordered by increasing recall.

Table 5 lists the retrievability inequality for a range of other rank cut-off factors on different queries sets. As expected, the Gini coefficient tends to decrease slowly for all query sets and models as the rank cut-off factor increases. The retrievability inequality within the collection is mitigated by the willingness of the user to search deeper down into the ranking. If users examine only the top documents, they will face a greater degree of retrieval bias. However, it is also obvious, that processing queries separately for both partitions greatly reduces the retrievability bias. Again, merging based on relative partition size performs better consistently.

## 5.2   Recall Evaluation for Prior-Art Search Task

The experiments above only considered retrievability as an abstract measure for evaluating bias in certain retrieval models, as well as a way to reduce that bias

**Table 7.** Number of prior art documents (Experiment D) classified as high and low retrievable

| Retr. Sys. | Low retrievable documents | % | High retrievable documents | % |
|---|---|---|---|---|
| TFIDF | 18659 | 54% | 15541 | 46% |
| BM25 | 14476 | 42% | 19724 | 58% |
| LM | 14745 | 43% | 19455 | 57% |

by applying it to a partitioned corpus. However, reducing retrievability bias does not automatically imply that this will also help us in improving accuracy within a certain rank cut-off for actual queries. In this section we thus analyze, in how far the approach of using a split document corpus will help increase the accuracy on actual prior-art search task patents.

To this end we select the 1,000 prior art query documents (Query Patents) of the TREC-CRT corpus. These documents have pointers to 34,200 prior art patents that have been identified as relevant prior art documents.

For each of the 1,000 query patents a set of 1- 2- 3- and 4-terms queries is created (exhaustive for the former two, a 20% limit for the later two). These are passed against the entire corpus of 1.2 million documents, using a rank cut-off of 150, 350, 550 and 750. For each document, the union of all result sets of queries created from these document is formed. The set is sorted according to the retrievability score, and the top-$c$ documents are returned as the ranked query result.

Table 7 lists the number of documents classified as high or low retrievable for the various retrieval models evaluated. Roughly half of all target documents are classified into the low-retrievable category. These are highly unlikely to be found using the query generation process employed for the underlying study if posed against a single corpus.

Table 6 lists the resulting accuracy values using the different retrieval models and merge strategies for a range of rank cut-off factors. While results definitely offer room for further improvement by considering document similarity measures during the merging stages, as well as by using more sophisticated query genera-tion algorithms. The results clearly show that overall retrievability is much higher using the retrieval via partitions size based merging, clearly outperforming the default strategy of performing retrieval on the entire corpus.

Overall, the default TFIDF retrieval model shows the worst accuracy. As we have seen in the initial retrievability experiments, it also exhibits the strongest retrievability inequality. This is a strong indicator that, at least for recall ori-ented applications, a strong retrieval bias (rendering many documents virtually unfindable) has a significant impact on retrieval performance. BM25 and LM perform almost identically as the recall on individual query patents depicted in Figure 5 shows.

The results above are not meant to be compared to accuracy rates of other retrieval engines operating on this corpus. They merely indicate that, by splitting a corpus into partitions of documents with high and low retrievability, gains in the overall retrieval accuracy can be observed. The exact amount of improvement achievable will depend both on the retrieval model as well as on the query generation process used.

## 6 Conclusions

This paper has introduced an approach to improve recall in recall oriented application domains by improving retrievability of documents. We first presented a detailed analysis of the characteristics of retrieval inequality in document corpora using a range of configurations for query types and retrieval models. We then introduced the concept of classifying documents automatically into high and low retrievable partitions using a set of simple features capturing term distribution characteristics. The classifier achieved classification accuracies in the range of 85%. By partitioning a corpus into documents that have potentially high and low retrievability, we are now able to perform retrieval separately on these two partitions, merging the result set afterwards. This increases the likelihood that even documents with low inherent retrievability can be found, leading to an overall higher accuracy.

While the results achieved are promising, several questions require more detailed evaluation. These include a more detailed analysis of the behavior of both retrievability as well as the improvement possible with more targeted retrieval system, rather than relying on the generation of exhaustive query sets. Furthermore, retrieval models may be optimized to exhibit minimal bias on the respective partitions, further improving retrievability and thus recall.

## References

1. Azzopardi, L., Vinay, V.: Retrievability: an evaluation measure for higher order information access tasks. In: CIKM '08: Proceeding of the 17th ACM Conference on Information and Knowledge Management, pp. 561–570. ACM, New York (2008)
2. Baeza-Yates, R.: Applications of web query mining. In: Losada, D.E., Fernández-Luna, J.M. (eds.) ECIR 2005. LNCS, vol. 3408, pp. 7–22. Springer, Heidelberg (2005)
3. Bashir, S., Rauber, A.: Analyzing document retrievability in patent retrieval settings. In: Bhowmick, S.S., Küng, J., Wagner, R. (eds.) DEXA 2009. LNCS, vol. 5690, pp. 753–760. Springer, Heidelberg (2009)
4. Bashir, S., Rauber, A.: Identification of low/high retrievable patents using content-based features. In: PaIR '09: Proceeding of the 2nd International Workshop on Patent Information Retrieval, pp. 9–16 (2009)
5. Custis, T., Al-Kofahi, K.: A new approach for evaluating query expansion: query-document term mismatch. In: SIGIR '07: Proceedings of the 30th Annual International ACM SIGIR Conference on Research and Development in Information Retrieval, pp. 575–582. ACM, New York (2007)
6. Doi, H., Seki, Y., Aono, M.: A patent retrieval method using a hierarchy of clusters at tut. In: NTCIR '05: In Proceedings of NTCIR-5 Workshop Meeting, Tokyo, Japan (December 6-9, 2005)

7. Fujii, A.: Enhancing patent retrieval by citation analysis. In: SIGIR '07: Proceedings of the 30th Annual International ACM SIGIR Conference on Research and Development in Information Retrieval, pp. 793–794. ACM, New York (2007)

8. Graf, E., Azzopardi, L.: A methodology for building a patent test collection for prior art search. In: EVIA '08: The Second International Workshop on Evaluating Information Access, Tokyo, Japan, pp. 60–71 (2008)

9. Itoh, H., Mano, H., Ogawa, Y.: Term distillation in patent retrieval. In: Proceedings of the ACL-2003 Workshop on Patent Corpus Processing, pp. 41–45. Association for Computational Linguistics (2003)

10. Jordan, C., Watters, C., Gao, Q.: Using controlled query generation to evaluate blind relevance feedback algorithms. In: JCDL '06: Proceedings of the 6th ACM/IEEE-CS Joint Conference on Digital Libraries, pp. 286–295. ACM, New York (2006)

11. Lupu, M., Huang, J., Zhu, J., Tait, J.: Trec-chem: large scale chemical information retrieval evaluation at trec. SIGIR Forum 43(2), 63–70 (2009)

12. Page, L., Brin, S., Motwani, R., Winograd, T.: The pagerank citation ranking: Bringing order to the web (1999)

13. Robertson, S., Zaragoza, H., Taylor, M.: Simple bm25 extension to multiple weighted fields. In: CIKM '04: Proceedings of the Thirteenth ACM International Conference on Information and Knowledge Management, pp. 42–49. ACM, New York (2004)

14. Robertson, S.E., Walker, S.: Some simple effective approximations to the 2-poisson model for probabilistic weighted retrieval. In: SIGIR '94: Proceedings of the 17th Annual International ACM SIGIR Conference on Research and Development in Information Retrieval, pp. 232–241. Springer, New York (1994)

15. Sakai, T.: Comparing metrics across trec and ntcir: the robustness to system bias. In: CIKM '08: Proceeding of the 17th ACM Conference on Information and Knowledge Management, pp. 581–590. ACM, New York (2008)

16. Vaughan, L., Thelwall, M.: Search engine coverage bias: evidence and possible causes. Inf. Process. Manage. 40(4), 693–707 (2004)

17. Witten, I.H., Frank, E.: Data mining: practical machine learning tools and techniques, 2nd edn. Morgan Kaufmann, USA (2005)

18. Xue, X., Croft, W.B.: Transforming patents into prior-art queries. In: SIGIR '09: Proceedings of the 32nd International ACM SIGIR Conference on Research and Development in Information Retrieval, pp. 808–809. ACM, New York (2009)

19. Zhai, C.: Risk minimization and language modeling in text retrieval dissertation abstract. SIGIR Forum 36(2), 100–101 (2002)

# Author Index

GPSR Compliance

The European Union's (EU) General Product Safety Regulation (GPSR)
is a set of rules that requires consumer products to be  safe and our
obligations to ensure this.

If you have any concerns about our products, you can contact us on
ProductSafety@springernature.com

In case Publisher is established outside the EU, the EU authorized
representative is:

Springer Nature Customer Service Center GmbH
Europaplatz 3
69115 Heidelberg, Germany


**Batch number: 09490872**


Printed by Printforce, the Netherlands